Buckle Down™

to the
Common Core
State Standards

Mathematics
Grade 7

This book belongs to: _____

Helping your schoolhouse meet the standards of the statehouse™

ISBN 978-0-7836-7989-1

1CCUS07MM01   1 2 3 4 5 6 7 8 9 10

Cover Image: Colorful marbles. © Corbis/Photolibrary

Triumph Learning® 136 Madison Avenue, 7th Floor, New York, NY 10016

© 2011 Triumph Learning, LLC
Buckle Down is an imprint of Triumph Learning®

Frequently Asked Questions about the Common Core State Standards

What are the Common Core State Standards?

The Common Core State Standards for mathematics and English language arts, grades K–12, are a set of shared goals and expectations for the knowledge and skills that will help students succeed. They allow students to understand what is expected of them and to become progressively more proficient in understanding and using mathematics and English language arts. Teachers will be better equipped to know exactly what they must do to help students learn and to establish individualized benchmarks for them.

Will the Common Core State Standards tell teachers how and what to teach?

No. Because the best understanding of what works in the classroom comes from teachers, these standards will establish *what* students need to learn, but they will not dictate *how* teachers should teach. Instead, schools and teachers will decide how best to help students reach the standards.

What will the Common Core State Standards mean for students?

The standards will provide a clear, consistent understanding of what is expected of student learning across the country. Common standards will not prevent different levels of achievement among students, but they will ensure more consistent exposure to materials and learning experiences through curriculum, instruction, teacher preparation, and other supports for student learning. These standards will help give students the knowledge and skills they need to succeed in college and careers.

Do the Common Core State Standards focus on skills and content knowledge?

Yes. The Common Core State Standards recognize that both content and skills are important. They require rigorous content and application of knowledge through higher-order thinking skills. The English language arts standards require certain critical content for all students, including classic myths and stories from around the world, America's founding documents, foundational American literature, and Shakespeare. The remaining crucial decisions about content are left to state and local determination. In addition to content coverage, the Common Core State Standards require that students systematically acquire knowledge of literature and other disciplines through reading, writing, speaking, and listening.

In mathematics, the Common Core State Standards lay a solid foundation in whole numbers, addition, subtraction, multiplication, division, fractions, and decimals. Together, these elements support a student's ability to learn and apply more demanding math concepts and procedures.

The Common Core State Standards require that students develop a depth of understanding and ability to apply English language arts and mathematics to novel situations, as college students and employees regularly do.

Will common assessments be developed?

It will be up to the states: some states plan to come together voluntarily to develop a common assessment system. A state-led consortium on assessment would be grounded in the following principles: allowing for comparison across students, schools, districts, states and nations; creating economies of scale; providing information and supporting more effective teaching and learning; and preparing students for college and careers.

TABLE OF CONTENTS

Common Core State Standards

**Common Core
State Standards**

To the Teacher:
Standards Name numbers are listed for each lesson in the table of contents. The numbers in the shaded gray bar that runs across the tops of the pages in the workbook indicate the Standards Name for a given page (see example to the left).

Introduction

Not a day goes by when you don't use your math skills in some way. You use math to figure out when to wake up in the morning so you can get to school on time, to decide whether you have enough money to buy a new DVD, and to estimate how much room is left on your computer's hard drive. You also use math when you check the temperature on a thermometer to decide whether you need to wear a jacket, and when you compare the amount of time you studied for a test to your score on the test.

This book will help you practice these and many other math skills that you can use in your everyday life, as well as in school. As with anything else, the more you practice these skills, the better you will get at applying them.

Test-Taking Tips

Here are a few tips that will help you on test day.

TIP 1: Take it easy.

Stay relaxed and confident. Because you've practiced these problems, you will be ready to do your best on almost any math test. Take a few slow, deep breaths before you begin the test.

TIP 2: Have the supplies you need.

For most math tests, you will need two sharp pencils and an eraser. Your teacher will tell you whether you need anything else.

TIP 3: Read the questions more than once.

Every question is different. Some questions are more difficult than others. If you need to, read a question more than once. This will help you make a plan for solving the question.

TIP 4: Learn to "plug in" answers to multiple-choice items.

When do you "plug in"? You should "plug in" whenever your answer is different from all of the answer choices or you can't come up with an answer. Plug each answer choice into the problem and find the one that makes sense. (You can also think of this as "working backward.")

TIP 5: Answer open-ended items completely.

When answering short-response and extended-response items, show all your work to receive as many points as possible. Write neatly enough so that your calculations will be easy to follow. Make sure your answer is clearly marked.

TIP 6: Use all the test time.

Work on the test until you are told to stop. If you finish early, go back through the test and double-check your answers. You just might increase your score on the test by finding and fixing any errors you might have made.

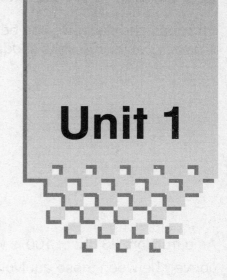

Unit 1

The Number System

Did you know that about $\frac{7}{10}$ of the Earth is covered by water, but only about 0.02 of the United States is covered by water? To work with facts like these, you need to use your number sense. Your brain gathers numbers of all sorts and processes them to make them understandable.

In this unit, you will review rational numbers. You will convert a rational number to a decimal, and compare and order rational numbers. You will add, subtract, multiply, and divide rational numbers. You will see that the properties of addition and multiplication that hold for integers also hold for rational numbers. Finally, you will solve problems involving rational numbers.

In This Unit

Rational Numbers

Addition and Subtraction of Rational Numbers

Properties of Addition

Multiplication and Division of Rational Numbers

Properties of Multiplication

Solving Problems with Rational Numbers

CCSS: 7.NS.2.b, 7.NS.2.d

Lesson 1: Rational Numbers

Fractions and decimals can be used to express parts of a whole. The following grid shows 43 of its 100 units shaded.

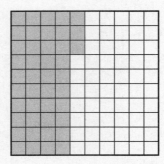

As a fraction, 43 out of 100 is written as $\frac{43}{100}$. As a decimal, it is written as 0.43. You can convert between these equivalent forms of the same quantity.

Fractions and decimals are forms of **rational numbers**. All rational numbers can be written as a fraction in which the numerator and denominator are integers, where the denominator is not 0. You can convert the fractional representation to a decimal by dividing the numerator by the denominator. This decimal expansion may be a **terminating decimal** or a **repeating decimal**.

You can convert a fraction to a decimal using long division. When you divide the numerator by the denominator, the resulting decimal will either terminate or repeat a digit or block of digits.

▶ Example

Convert $\frac{1}{8}$ to a decimal.

$$
\begin{array}{r}
0.125 \\
8\overline{)1.000} \\
-8 \\
\hline
20 \\
-16 \\
\hline
40 \\
-40 \\
\hline
0 \text{ (terminating)}
\end{array}
$$

$\frac{1}{8} = 0.125$

 TIP: Recall that **integers** are the set of all whole numbers and their opposites: ..., −3, −2, −1, 0, 1, 2, 3, ...

CCSS: 7.NS.2.b, 7.NS.2.d

 Example

Convert $\frac{2}{9}$ to a decimal.

$$\begin{array}{r} 0.222 \\ 9\overline{)2.000} \\ -18 \\ \hline 20 \\ -18 \\ \hline 20 \\ -18 \\ \hline 2 \text{ (repeating)} \end{array}$$

$\frac{2}{9} = 0.222\ldots$

TIP: You can write a bar, —, over one or more digits to show that they repeat. For example, $\frac{2}{9} = 0.22\ldots = 0.\overline{2}$ and $\frac{3}{11} = 0.272727\ldots = 0.\overline{27}$.

An integer can be expressed as a fraction by writing the integer in the numerator and writing 1 in the denominator.

 Example

Write 4 as a fraction.

Write the integer over a denominator of 1 to express it as a fraction.
$4 = \frac{4}{1}$

4 written as a fraction is $\frac{4}{1}$.

 Example

Write -7 as a fraction.

Write the integer over a denominator of 1 to express it as a fraction.
$-7 = \frac{-7}{1}$

-7 written as a fraction is $\frac{-7}{1}$.

CCSS: 7.NS.2.b, 7.NS.2.d

To compare different forms of rational numbers, first convert all of them to the same form.

 Example

Compare $\frac{3}{4}$ and 0.78.

Convert $\frac{3}{4}$ to a decimal.

$$
\begin{array}{r}
0.75 \\
4\overline{)3.00} \\
-2\,8 \\
\hline
20 \\
-20 \\
\hline
0
\end{array}
$$

$\frac{3}{4} = 0.75$

Use $<$, $>$, or $=$ to compare the decimals.
$0.75 < 0.78$

Therefore, $\frac{3}{4} < 0.78$.

 Example

Compare $3\frac{2}{5}$ and 3.25.

Convert $3\frac{2}{5}$ to a decimal.

$$
\begin{array}{r}
0.4 \\
5\overline{)2.0} \\
-2\,0 \\
\hline
0
\end{array}
$$

$3\frac{2}{5} = 3.4$

Use $<$, $>$, or $=$ to compare the decimals.
$3.4 > 3.25$

Therefore, $3\frac{2}{5} > 3.25$.

 TIP: To change a mixed number to a decimal, convert only the fraction part to a decimal. Leave the whole number part the same, and include the decimal value.

CCSS: 7.NS.2.b, 7.NS.2.d

You can use a number line to order rational numbers. On a number line, a number is less than any numbers to its right and greater than any numbers to its left.

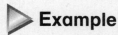 **Example**

Order the numbers from least to greatest.

$$1.5 \qquad -1\frac{7}{8} \qquad \frac{4}{5} \qquad -0.4$$

Plot and label each point on a number line.

1.5 is halfway between 1 and 2. Put a dot at 1.5.

$-1\frac{7}{8}$ is between -2 and -1, but is closer to -2. Put a dot at $-1\frac{7}{8}$.

$\frac{4}{5}$ is between 0 and 1, but is closer to 1. Put a dot at $\frac{4}{5}$.

-0.4 is between -1 and 0, but is closer to 0. Put a dot at -0.4.

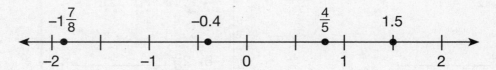

Read the numbers on the number line from left to right.

The numbers from least to greatest are $-1\frac{7}{8}$, -0.4, $\frac{4}{5}$, 1.5.

Practice

Directions: For questions 1 through 8, find the decimal form of the given number.

1. $\frac{-1}{4}$ _____

2. $\frac{1}{2}$ _____

3. $\frac{5}{8}$ _____

4. $\frac{9}{11}$ _____

5. $\frac{2}{3}$ _____

6. $\frac{11}{16}$ _____

7. $1\frac{7}{20}$ _____

8. $2\frac{5}{9}$ _____

9. How is $\frac{24}{27}$ written as a decimal?

A. 0.8

B. $0.\overline{8}$

C. 0.9

D. 1.125

10. How is $\frac{12}{16}$ written as a decimal?

A. $0.\overline{3}$

B. 0.34

C. 0.75

D. $1.\overline{3}$

CCSS: 7.NS.2.b, 7.NS.2.d

11. Until recently, share prices at the New York Stock Exchange were displayed using fractions. What is the equivalent decimal value of a stock whose share price was 4\frac{7}{8}$?

_____

12. Timothy bought 9 tomatoes for $4 at a farm stand. What is the price per tomato in decimal form? Round your answer to the nearest hundredth.

_____

13. While the currency of the United States is dollars, the currency of England is pounds. When Robin visited London, she exchanged $225 for 150 pounds. What was the exchange rate of pounds to dollars in decimal form? Round your answer to the nearest hundredth.

_____

14. Javier worked for 16 hours last week and was paid $157. How much did Javier earn per hour to the nearest cent?

_____

CCSS: 7.NS.2.b, 7.NS.2.d

Directions: For questions 15 through 26, use $<$, $>$, or $=$ to compare the numbers.

15. $8\frac{4}{5}$ _____ 8.08

16. 9.95 _____ $9\frac{10}{11}$

17. 0.52 _____ $\frac{13}{25}$

18. 1.35 _____ $1\frac{3}{5}$

19. $\frac{3}{8}$ _____ $0.3\overline{7}$

20. $\frac{1}{2}$ _____ $0.\overline{5}$

21. 0.015 _____ $\frac{3}{20}$

22. 0.67 _____ $\frac{19}{25}$

23. $\frac{-18}{25}$ _____ $-0.\overline{72}$

24. $3\frac{9}{10}$ _____ 3.09

25. $\frac{29}{6}$ _____ 4.85

26. $5\frac{1}{7}$ _____ 5.17

CCSS: 7.NS.2.b, 7.NS.2.d

Directions: For questions 27 through 30, plot and label each point on the number line.

27. $2\frac{2}{5}$ 1.87 -1 $\frac{3}{4}$

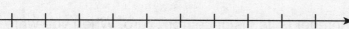

28. $\frac{-1}{2}$ -1.2 $\frac{7}{3}$ 0.44

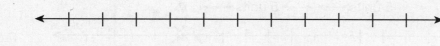

29. $\frac{-1}{4}$ $\frac{5}{8}$ 1.63 -2

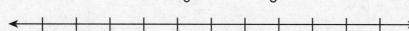

30. -0.8 1.35 $\frac{13}{6}$ $-1\frac{7}{8}$

31. Geraldine is doing an experiment to study plant growth. She measured the heights of her plants to be $\frac{5}{6}$ ft, 0.85 ft, $\frac{3}{4}$ ft, and 0.9 ft. List the heights of these plants in order from least to greatest.

_____

32. Jimmy measured four pieces of wood. Their measures are 4.2 ft, $\frac{9}{2}$ ft, 4.02 ft, and $4\frac{2}{9}$ ft.

What is the length of the shortest piece?

_____

What is the length of the longest piece?

_____

Explain how you found your answers.

_____

_____

_____

CCSS: 7.NS.1.a, 7.NS.1.b, 7.NS.1.c, 7.NS.3

Lesson 2: Addition and Subtraction of Rational Numbers

The **absolute value** of an integer is its distance from zero. When you write the absolute value of an integer, *n*, use the notation |*n*|. The following number line shows that −5 and 5 are both 5 units from zero. Therefore, |−5| and |5| are both equal to 5.

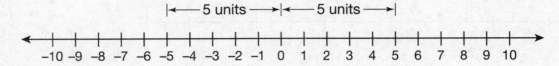

The sum of *p* + *q* is the number located a distance |*q*| from *p* in the positive or negative direction, depending whether *q* is positive or negative.

 Example

Find the sum of 0.7 + 0.4.

Use a number line. For 0.7 + 0.4, *p* = 0.7 and *q* = 0.4. |*q*| = |0.4| = 0.4. Since *q* is positive, find the number that is 0.4 units to the right of 0.7.

0.7 + 0.4 = 1.1

 Example

Find the sum of $\frac{3}{8} + \left(-\frac{7}{8}\right)$.

Use a number line. For $\frac{3}{8} + \left(-\frac{7}{8}\right)$, $p = \frac{3}{8}$ and $q = -\frac{7}{8}$. $|q| = \left|-\frac{7}{8}\right| = \frac{7}{8}$.

Since *q* is negative, find the number that is $\frac{7}{8}$ units to the left of $\frac{3}{8}$.

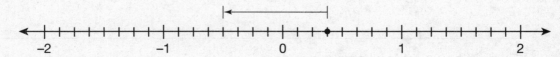

$\frac{3}{8} + \left(-\frac{7}{8}\right) = -\frac{4}{8}$

You can simplify $-\frac{4}{8}$ to $-\frac{1}{2}$.

$\frac{3}{8} + \left(-\frac{7}{8}\right) = -\frac{1}{2}$

CCSS: 7.NS.1.a, 7.NS.1.b, 7.NS.1.c, 7.NS.3

 Example

Find the sum of $-\frac{9}{10} + \left(-\frac{4}{10}\right)$.

Use a number line. For $-\frac{9}{10} + \left(-\frac{4}{10}\right)$, $p = -\frac{9}{10}$ and $q = -\frac{4}{10}$.
$|q| = \left|-\frac{4}{10}\right| = \frac{4}{10}$. Since q is negative, find the number that is $\frac{4}{10}$ units to the left of $-\frac{9}{10}$.

$-\frac{9}{10} + \left(-\frac{4}{10}\right) = -\frac{13}{10} = -1\frac{3}{10}$

Two numbers are **opposites** if they are the same distance from 0 on a number line, but in opposite directions. The numbers 4 and −4 are opposites. Opposites are also called **additive inverses**. A number and its opposite have a sum of 0.

 Example

Ella had $50 in her bank account. She then withdrew $50 from her account. How much does Ella have in her bank account now?

Represent the numbers with integers.
$50 in her bank account can be represented by 50.
Withdrawing $50 can be represented by −50.

$50 + (-50) = 0$

Ella has $0 in her bank account now.

Subtracting an integer gives the same result as adding its additive inverse.
$p - q = p + (-q)$

▶ **Example**

Find $-2 - 7$.

Use a number line. Rewrite $-2 - 7$ as $-2 + (-7)$. Find the number that is 7 units to the left of −2.

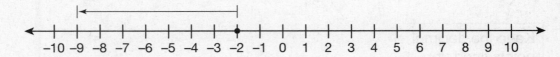

$-2 - 7 = -9$

19

The distance between two numbers on a number line is equal to the absolute value of their difference.

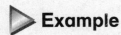 **Example**

What is the distance between −5 and 4?

Use a number line.

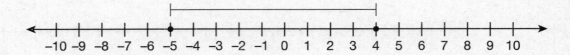

Count the distance on the number line. The distance between −5 and 4 is 9 units.

Check by finding the absolute value of the difference of the numbers.

$$|-5 - 4|$$
$$= |-5 + (-4)|$$
$$= |-9|$$
$$= 9$$

$$|4 - (-5)|$$
$$= |4 + (+5)|$$
$$= |9|$$
$$= 9$$

The following examples review computation with decimals, fractions, and mixed numbers.

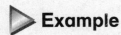 **Examples**

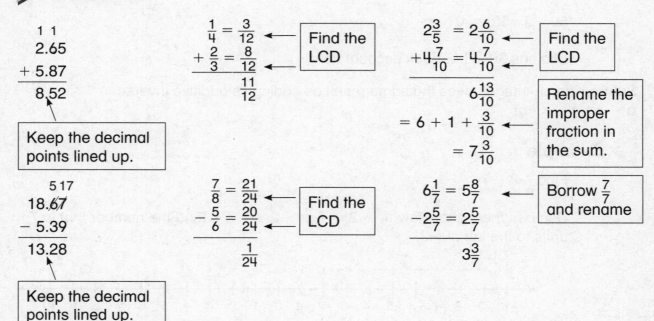

$$\begin{array}{r} 1\ 1 \\ 2.65 \\ +\ 5.87 \\ \hline 8.52 \end{array}$$

Keep the decimal points lined up.

$$\frac{1}{4} = \frac{3}{12}$$
$$+\frac{2}{3} = \frac{8}{12}$$
$$\overline{\frac{11}{12}}$$

Find the LCD

$$2\frac{3}{5} = 2\frac{6}{10}$$
$$+4\frac{7}{10} = 4\frac{7}{10}$$
$$\overline{6\frac{13}{10}}$$
$$= 6 + 1 + \frac{3}{10}$$
$$= 7\frac{3}{10}$$

Find the LCD

Rename the improper fraction in the sum.

$$\begin{array}{r} 5\ 17 \\ 18.6\!\!\!/7 \\ -\ 5.39 \\ \hline 13.28 \end{array}$$

Keep the decimal points lined up.

$$\frac{7}{8} = \frac{21}{24}$$
$$-\frac{5}{6} = \frac{20}{24}$$
$$\overline{\frac{1}{24}}$$

Find the LCD

$$6\frac{1}{7} = 5\frac{8}{7}$$
$$-2\frac{5}{7} = 2\frac{5}{7}$$
$$\overline{3\frac{3}{7}}$$

Borrow $\frac{7}{7}$ and rename

▶ **Example**

Fort Lauderdale, FL, is about 9 feet above sea level and New Orleans, LA, is about 6 feet below sea level. What is the difference in their elevations?

Represent the numbers with integers.
9 feet above sea level can be represented by 9.
6 feet below sea level can be represented by -6.

$9 - (-6) = 9 + (+6) = 15$ The difference in the elevations is 15 feet.

▶ **Example**

After school Brandon walked $\frac{3}{4}$ mile to his friend Simone's house. Then he walked $\frac{2}{3}$ mile from Simone's house to his own house. How far did Brandon walk in all after school?

$$\frac{3}{4} = \frac{9}{12}$$
$$+\frac{2}{3} = \frac{8}{12}$$
$$\frac{17}{12} = 1\frac{5}{12}$$

Brandon walked $1\frac{5}{12}$ miles in all after school.

▶ **Example**

The Wilsons drove 954.8 miles from Louisville, KY, to Houston, TX. Then they drove another 570.6 miles from Houston to Memphis, TN. How far did they drive in all?

$$954.8$$
$$+\ 570.6$$
$$\overline{1525.4}$$

The Wilsons drove 1525.4 miles.

▶ **Example**

Huan removed $8\frac{2}{3}$ yards of material from a roll that contained $24\frac{1}{2}$ yards of the material. How much material was left on the roll?

$$24\frac{1}{2} = 24\frac{3}{6} = 23\frac{9}{6}$$
$$-8\frac{2}{3} = \ 8\frac{4}{6} = \ 8\frac{4}{6}$$
$$\overline{15\frac{5}{6}}$$

There were $15\frac{5}{6}$ yards of material left on the roll.

CCSS: 7.NS.1.a, 7.NS.1.b, 7.NS.1.c, 7.NS.3

⬤ Practice

Directions: For questions 1 through 4, use the number line. Simplify, when possible.

1. $0.3 + 0.9$

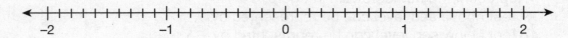

_____

2. $\frac{1}{5} + \left(\frac{-4}{5}\right)$

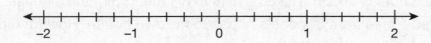

_____

3. $\frac{-1}{8} + \left(\frac{-5}{8}\right)$

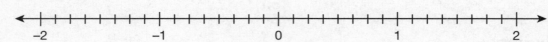

_____

4. $-6 + 6$

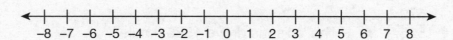

_____

CCSS: 7.NS.1.a, 7.NS.1.b, 7.NS.1.c, 7.NS.3

Directions: For questions 5 through 24, compute. Write answers in simplest form.

5. $-12 + 19$ _____

6. $-0.7 + 2.1$ _____

7. $\dfrac{-5}{9} + \left(\dfrac{-2}{9}\right)$ _____

8. $\dfrac{11}{12} + \left(\dfrac{-11}{12}\right)$ _____

9. $3.6 + (-1.7)$ _____

10. $-4.5 + (-9.6)$ _____

11. $-26.3 + 26.3$ _____

12. $\dfrac{7}{8} + \left(\dfrac{-3}{4}\right)$ _____

13. $-15 - (-8)$ _____

14. $\dfrac{2}{3} - \left(\dfrac{-1}{3}\right)$ _____

15. $4\dfrac{4}{5} + 3\dfrac{2}{5}$ _____

16. $\dfrac{19}{20} - \dfrac{1}{2}$ _____

17. $11\dfrac{8}{9} - 6\dfrac{2}{9}$ _____

18. $33.07 - 17.54$ _____

19. $6\dfrac{1}{3} - 4\dfrac{5}{6}$ _____

20. $-49 - (-49)$ _____

21. $4\dfrac{7}{8} + 9\dfrac{1}{3}$ _____

22. $-15.8 - 15.8$ _____

23. $8\dfrac{3}{4} - 10\dfrac{2}{3}$ _____

24. $-4.85 - 7.6$ _____

25. Henry has $75 in his bank account. If he withdraws $75 from the account, how much money will be left in the account?

_____

26. A movie is advertised to last for $2\frac{1}{3}$ hours. Of this time, $\frac{1}{4}$ hour is used for showing previews. How long is the actual movie?

_____

27. Lily finished her first lap of a 200-meter race in 14.76 seconds and her second lap in 15.17 seconds. What was her total time for the two laps?

_____

28. On a cold winter day, the low temperature was 10°F. Due to the wind chill, it felt like −18°F. What was the difference between the actual temperature and the wind chill temperature?

_____

29. Last night Jeremy studied $\frac{5}{6}$ hour for his math test and $\frac{1}{4}$ hour for his spelling test. How long did Jeremy study altogether for the two tests?

_____

30. In a race at the track meet, Rosa finished in 7.5 minutes and Chad finished in 8.25 minutes. How much more time did it take Chad to finish the race than Rosa?

_____

CCSS: 7.NS.1.a, 7.NS.1.b, 7.NS.1.c, 7.NS.3

31. A submarine is at a depth of 400 meters below sea level. The submarine then rises by 185 meters. What is the depth of the submarine now?

    A. 215 meters above sea level

    B. 215 meters below sea level

    C. 585 meters below sea level

    D. 585 meters above sea level

32. In 1935, Jesse Owens set a long jump world record of 8.13 meters. In 1968 Bob Beamon surpassed Jesse Owens by jumping 8.9 meters. How much farther did Bob Beamon jump than Jesse Owens?

    A. 0.04 meter

    B. 0.4 meter

    C. 0.77 meter

    D. 0.87 meter

33. The temperature at noon was 6°F. At midnight it was −4°F.

    What was the difference in temperatures?

    _____

    Explain how you found your answer.

    _____

    _____

    _____

    Explain how you could use a number line to find the difference in temperatures.

    _____

    _____

    _____

Lesson 3: Properties of Addition

A number property states a relationship between numbers. Number properties of addition are true for all rational numbers, including integers, fractions, and decimals.

The **associative property of addition** states that the way in which numbers are grouped when more than two numbers are added does not matter.

$$(a + b) + c = a + (b + c)$$

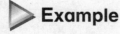 **Example**

Show that $(6.3 + 3.8) + 5.9 = 6.3 + (3.8 + 5.9)$ is a true statement.
$$(6.3 + 3.8) + 5.9 = 6.3 + (3.8 + 5.9)$$
$$10.1 + 5.9 = 6.3 + 9.7$$
$$16 = 16$$

The **commutative property of addition** states that the order in which numbers are added does not matter.

$$a + b = b + a$$

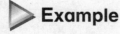 **Example**

Show that $\frac{4}{15} + \frac{7}{15} = \frac{7}{15} + \frac{4}{15}$ is a true statement.
$$\frac{4}{15} + \frac{7}{15} = \frac{7}{15} + \frac{4}{15}$$
$$\frac{11}{15} = \frac{11}{15}$$

The **additive identity property of 0** states that the sum of any number and 0 is equal to that number.

$$a + 0 = 0 + a = a$$

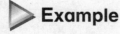 **Example**

Show that $-8 + 0 = 0 + (-8)$ is a true statement.
$$-8 + 0 = 0 + (-8)$$
$$-8 = -8$$

CCSS: 7.NS.1.d

 Example

What number makes $\left(8\frac{3}{5} + 2\frac{2}{3}\right) + 3\frac{1}{3} = 8\frac{3}{5} + \left(\square + 3\frac{1}{3}\right)$ a true statement?

Use the associative property of addition.

$$\left(8\frac{3}{5} + 2\frac{2}{3}\right) + 3\frac{1}{3} = 8\frac{3}{5} + \left(2\frac{2}{3} + 3\frac{1}{3}\right)$$

The number $2\frac{2}{3}$ makes $\left(8\frac{3}{5} + 2\frac{2}{3}\right) + 3\frac{1}{3} = 8\frac{3}{5} + \left(\square + 3\frac{1}{3}\right)$ a true statement.

You can use the properties of addition to help you simplify expressions and find sums.

 Example

What is $(-13 + 0) + 9$?

Use the additive identity property of 0.
$(-13 + 0) + 9 = -13 + 9 = -4$

$(-13 + 0) + 9 = -4$

 Example

What is $(7.5 + 4.7) + 2.5$?

Use the commutative property of addition to reorder the addends in the parentheses.
$(7.5 + 4.7) + 2.5 = (4.7 + 7.5) + 2.5$

Use the associative property of addition to regroup the addends.
$(4.7 + 7.5) + 2.5 = 4.7 + (7.5 + 2.5)$

Add.
$4.7 + (7.5 + 2.5) = 4.7 + 10 = 14.7$

$(7.5 + 4.7) + 2.5 = 14.7$

TIP: The commutative and associative properties do not hold for subtraction.

⬤ Practice

Directions: For questions 1 through 6, write the property that is represented by the given equation.

1. $0 + \frac{7}{12} = \frac{7}{12}$ _____

2. $5.4 + 2.9 = 2.9 + 5.4$ _____

3. $\left(9\frac{1}{2} + 3\frac{3}{4}\right) + 4\frac{1}{4} = 9\frac{1}{2} + \left(3\frac{3}{4} + 4\frac{1}{4}\right)$ _____

4. $-6.1 + 0 = -6.1$ _____

5. $\frac{-9}{10} + \frac{1}{2} = \frac{1}{2} + \left(\frac{-9}{10}\right)$ _____

6. $8.2 + (4.9 + 3.8) = (8.2 + 4.9) + 3.8$ _____

Directions: For questions 7 through 10, complete each equation to make it a true statement.

7. $(5.5 + 7.1) + 6.5 = 5.5 + (7.1 + $ _____$)$

8. $-21 + $ _____ $= -21$

9. $\frac{3}{8} + \frac{7}{8} = $ _____ $+ \frac{3}{8}$

10. $\left(2\frac{3}{10} + 7\frac{1}{5}\right) + 9\frac{7}{10} = 2\frac{3}{10} + \left($ _____ $+ 9\frac{7}{10}\right)$

11. Use the commutative property of addition to complete the equation. Then simplify to show that the two sides of the equation are equal.

    $\frac{7}{12} + \frac{11}{12} =$ _____

12. Use the associative property of addition to complete the equation. Then simplify to show that the two sides of the equation are equal.

    $(3.7 + 6.5) + 4.5 =$ _____

13. What is the value of p in this equation?

    $\frac{4}{5} + \left(p + 3\frac{1}{2}\right) = \left(\frac{4}{5} + \frac{1}{5}\right) + 3\frac{1}{2}$

    A. $\frac{1}{5}$

    B. $\frac{4}{5}$

    C. 1

    D. $3\frac{1}{2}$

14. Which statement is true?

    A. $-7 + (-11) = -11 + 7$

    B. $\left(0 + \frac{5}{6}\right) = \frac{1}{6}$

    C. $\left(3\frac{7}{8} + 5\frac{1}{2}\right) + 8\frac{1}{2} = 3\frac{7}{8} + \left(5\frac{1}{2} + 8\frac{1}{2}\right)$

    D. $9.4 + 1 = 9.4$

15. Fill in the missing information to find the sum of $5\frac{1}{12} + \left(8\frac{1}{2} + 6\frac{11}{12}\right)$.

    $5\frac{1}{12} + \left(8\frac{1}{2} + 6\frac{11}{12}\right) = 5\frac{1}{12} + \left(6\frac{11}{12} + 8\frac{1}{2}\right)$ by the _____ property of addition.

    $5\frac{1}{12} + \left(6\frac{11}{12} + 8\frac{1}{2}\right) = \left(5\frac{1}{12} + 6\frac{11}{12}\right) + 8\frac{1}{2}$ by the _____ property of addition.

    $\left(5\frac{1}{12} + 6\frac{11}{12}\right) + 8\frac{1}{2} =$ _____ $+ 8\frac{1}{2} =$ _____

16. Fill in the missing information to find the sum of $(-13 + 27) + 13$.

    $(-13 + 27) + 13 = [27 + (-13)] + 13$ by the _____ property of addition.

    $[27 + (-13)] + 13 = 27 + (-13 + 13)$ by the _____ property of addition.

    $27 + (-13 + 13) = 27 +$ _____

    $27 +$ _____ $=$ _____ by the _____ property of 0.

CCSS: 7.NS.2.a, 7.NS.2.b, 7.NS.3

Lesson 4: Multiplication and Division of Rational Numbers

The following examples review computation with decimals, fractions, and mixed numbers.

 Examples

$$\begin{array}{r} 8.1 \\ \times\ 2.6 \\ \hline 4\ 8\ 6 \\ 16\ 2\ 0 \\ \hline 21.0\ 6 \end{array}$$

$$\frac{5}{9} \cdot \frac{3}{4} = \frac{15}{36}$$

$$= \frac{5}{12}$$

Simplify to lowest terms.

$$1\frac{2}{3} \cdot 1\frac{1}{5} = \frac{5}{3} \cdot \frac{6}{5}$$

$$= \frac{30}{15}$$

$$= 2$$

Convert to improper fractions.

Move the decimal point 2 places to the left.

$$\frac{1}{8} \div \frac{3}{4} = \frac{1}{8} \cdot \frac{4}{3}$$

$$= \frac{4}{24}$$

$$= \frac{1}{6}$$

Multiply by the reciprocal.

$$3\frac{1}{4} \div 1\frac{1}{2} = \frac{13}{4} \div \frac{3}{2}$$

$$= \frac{13}{4} \cdot \frac{2}{3}$$

$$= \frac{26}{12}$$

$$= \frac{13}{6}$$

$$= 2\frac{1}{6}$$

Convert to improper fractions.

Keep the decimal points lined up.

$$\begin{array}{r} 0.86 \\ 3\overline{)2.58} \\ -2\ 4 \\ \hline 18 \\ -\ 18 \\ \hline 0 \end{array}$$

 TIP: In a **complex fraction**, both the numerator and denominator are fractions.

$\dfrac{\frac{3}{4}}{\frac{5}{6}}$ is a complex fraction. To simplify a complex fraction, divide the numerator by the denominator.

$$\frac{\frac{3}{4}}{\frac{5}{6}} = \frac{3}{4} \div \frac{5}{6} = \frac{3}{4} \cdot \frac{6}{5} = \frac{18}{20} = \frac{9}{10}$$

CCSS: 7.NS.2.a, 7.NS.2.b, 7.NS.3

Follow these rules for multiplying or dividing two integers.

The product or quotient of two positive integers is positive.

▶ **Example**

$$6 \cdot 7 = 42$$
$$24 \div 4 = 6$$

The product or quotient of two negative integers is positive.

▶ **Example**

$$-4 \cdot -2 = 8$$
$$-36 \div -9 = 4$$

The product or quotient of two integers with different signs (one positive and one negative) is negative.

▶ **Example**

$$3 \cdot -4 = -12$$
$$-35 \div 7 = -5$$

Every quotient of integers is a rational number.

If p and q are integers, then $-\left(\dfrac{p}{q}\right) = \dfrac{(-p)}{q} = \dfrac{p}{(-q)}$, where q is not 0.

▶ **Examples**

$$-\frac{7}{8} = \frac{-7}{8} = \frac{7}{-8}$$

$$\frac{27}{-9} = \frac{-27}{9} = -\frac{27}{9} = -3$$

31

The rules for multiplying and dividing integers also apply to rational numbers.

▷ **Example**

Multiply: $-\frac{2}{5} \cdot \frac{3}{4}$

Multiply the fractions.

$$\frac{2}{5} \cdot \frac{3}{4} = \frac{6}{20}$$
$$= \frac{3}{10}$$

The product of two rational numbers with different signs (one positive and one negative) is negative.

Therefore, $-\frac{2}{5} \cdot \frac{3}{4} = -\frac{3}{10}$.

When dividing decimals, move the decimal point to the right in the divisor so that the divisor is a whole number. Move the decimal point the same number of places to the right in the dividend. Remember to move the decimal point straight up in your answer.

▷ **Example**

Divide: $-59.36 \div -1.6$

Move the decimal point one place to the right in both the divisor and the dividend to make the divisor a whole number. Divide, and then place the decimal point straight up in the quotient.

$$
1.6\overline{)59.36} \rightarrow
\begin{array}{r}
37.1 \\
16\overline{)593.6} \\
-48 \\
\hline
113 \\
-112 \\
\hline
16 \\
-16 \\
\hline
0
\end{array}
$$

The product or quotient of two negative rational numbers is positive.

Therefore, $-59.36 \div -1.6 = 37.1$.

CCSS: 7.NS.2.a, 7.NS.2.b, 7.NS.3

▶ Example

From 6 P.M. until midnight, the temperature change was $-12°F$. What was the average change in temperature per hour?

From 6 P.M. until midnight is 6 hours.
$-12 \div 6 = -2$

The average change per hour was $-2°F$.

▶ Example

The Seven Sisters cliffs in England are retreating at a rate of about 0.35 meter each year. About how much do the cliffs retreat over 5 years?

Represent the rate of retreat each year with -0.35.

$-0.35 \cdot 5 = -1.75$

-1.75 represents the amount the cliffs retreat.

The cliffs retreat about 1.75 meters over 5 years.

▶ Example

A can of corn contains $1\frac{1}{3}$ cups of corn. The suggested serving size is $\frac{1}{3}$ cup. How many servings of corn are in the can?

$$
\begin{aligned}
1\frac{1}{3} \div \frac{1}{3} &= \frac{4}{3} \div \frac{1}{3} \\
&= \frac{4}{3} \cdot \frac{3}{1} \\
&= \frac{12}{3} \\
&= 4
\end{aligned}
$$

There are 4 servings of corn in the can.

▶ Example

A diver descends at a rate of 15 feet per minute. What integer represents the depth of the diver below the surface after 7 minutes?

Represent the rate with an integer. Descending at 15 feet per minute can be represented by -15.

$-15 \cdot 7 = -105$

-105 represents 105 feet below the surface.

The diver will be at a depth of 105 feet below the surface after 7 minutes.

CCSS: 7.NS.2.a, 7.NS.2.b, 7.NS.3

 Practice

Directions: For questions 1 through 16, carry out the given operation. Write answers in simplest form.

1. $\frac{-52}{4}$ _____

2. $-15 \cdot -8$ _____

3. $2\frac{2}{5} \cdot \frac{5}{12}$ _____

4. $\frac{5}{6} \div \frac{8}{9}$ _____

5. $10.4 \cdot 0.89$ _____

6. $58.09 \div 7.4$ _____

7. $-48 \div -6$ _____

8. $-15.2 \cdot 3.6$ _____

9. $1\frac{5}{6} \cdot 12$ _____

10. $4.41 \div 9$ _____

11. $47.75 \div 0.25$ _____

12. $-64 \cdot \frac{5}{8}$ _____

13. $3\frac{3}{5} \div \frac{9}{10}$ _____

14. $\frac{2}{3} \cdot 2\frac{1}{4}$ _____

15. $2\frac{4}{5} \div 3\frac{1}{2}$ _____

16. $3.4 \cdot -0.15$ _____

17. Multiply: $-8 \cdot -2\frac{3}{5}$

    A. $2\frac{4}{5}$ C. $21\frac{4}{5}$

    B. $20\frac{4}{5}$ D. 28

18. Which statement is true if a and b are integers, $a \neq b$, and neither a nor b is equal to 0?

    A. $\frac{-a}{b} = \frac{-b}{a}$ C. $\frac{-a}{-b} = -\frac{a}{b}$

    B. $-\frac{a}{b} = \frac{a}{b}$ D. $\frac{-a}{b} = \frac{a}{-b}$

19. What is $-3 \cdot -5 \cdot -2$?

    A. -30 C. 10

    B. -10 D. 30

20. Barney forgot his lunch money 5 days in a row. Each day he borrowed $4 from Kevin. Which of the following represents how much less money Kevin has now?

    A. $4 \cdot -5 = -20$ C. $-4 \cdot 5 = -20$

    B. $-4 \cdot -5 = 20$ D. $-4 + 5 = 1$

21. Jamal bought $3\frac{3}{4}$ pounds of trail mix, which he divided among several bags. Each bag contained $\frac{3}{8}$ pound of trail mix. How many bags of trail mix did Jamal make?

    A. 4   C. 8

    B. 5   D. 10

22. A submarine at a depth of 675 feet below the surface rose to $\frac{1}{3}$ this depth. Which of the following represents the new depth of the submarine below the surface?

    A. 225 feet   C. $675\frac{1}{3}$ feet

    B. $674\frac{2}{3}$ feet   D. 2025 feet

23. Over a 5-day period, the value of a stock had a total change of -4.35. What was the average change per day?

_____

24. Fred has a bookcase that is 41.5 inches high. If shelves are evenly-spaced and about 8.3 inches apart, how many shelves are on Fred's bookcase?

_____

25. It took Rachel 9.3 minutes to run a mile. It took Chris 0.9 times as long as it took Rachel to run the same distance. How long did it take Chris to run a mile?

_____

26. A bag of corn chips holds $12\frac{1}{2}$ ounces. Each serving of corn chips is $1\frac{1}{4}$ ounces. How many servings are in each bag of corn chips?

_____

27. Allyson has 4.5 pieces of rope to use for a tire swing. If each full piece is $2\frac{2}{3}$ feet long, how much rope does Allyson have in all?

_____

28. If the two factors in a multiplication problem are integers with different signs, how does their product compare to either number? Give an example to support your answer.

_____

_____

29. There is an error in the following statement. Rewrite the statement to correct it.

    If an integer is divided by an integer with the opposite sign, the quotient has the sign of the integer with the greater absolute value.

_____

_____

_____

30. The table below shows the number of degrees that the temperature changed over each hour for 4 hours.

Hour	1	2	3	4
Change in Temperature °F	−3	−1	−2	−6

What was the total change in temperature? _____

Explain how you found your answer.

_____

_____

What was the average change in temperature per hour?

_____

Explain how you found your answer.

_____

_____

CCSS: 7.NS.2.a, 7.NS.2.c

Lesson 5: Properties of Multiplication

Number properties of multiplication work for all rational numbers, including integers, fractions, and decimals.

The **associative property of multiplication** states that the grouping of terms that are multiplied together does not change the product.

$$(a \bullet b) \bullet c = a \bullet (b \bullet c)$$

▷ **Example**

Show that $\left(\frac{2}{3} \bullet \frac{1}{3}\right) \bullet 18 = \frac{2}{3} \bullet \left(\frac{1}{3} \bullet 18\right)$ is a true statement.

$$\left(\frac{2}{3} \bullet \frac{1}{3}\right) \bullet 18 = \frac{2}{3} \bullet \left(\frac{1}{3} \bullet 18\right)$$

$$\frac{2}{9} \bullet 18 = \frac{2}{3} \bullet 6$$

$$4 = 4$$

The **commutative property of multiplication** states that the order in which terms are multiplied does not change the product.

$$a \bullet b = b \bullet a$$

▷ **Example**

Show that $1.5 \bullet 2.8 = 2.8 \bullet 1.5$ is a true statement.

$$1.5 \bullet 2.8 = 2.8 \bullet 1.5$$

$$4.2 = 4.2$$

The **multiplicative identity property of 1** states that the product of any number and 1 is equal to that number.

$$a \bullet 1 = 1 \bullet a = a$$

▷ **Example**

Show that $-7 \bullet 1 = 1 \bullet (-7)$ is a true statement.

$$-7 \bullet 1 = 1 \bullet (-7)$$

$$-7 = -7$$

CCSS: 7.NS.2.a, 7.NS.2.c

The **distributive property of multiplication** relates multiplication to addition or subtraction. The property states that everything inside the parentheses is multiplied by whatever is outside the parentheses.

$$a \bullet (b + c) = a \bullet b + a \bullet c$$

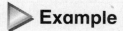 **Example**

Show that $\frac{1}{2} \bullet (24 + 38) = \frac{1}{2} \bullet 24 + \frac{1}{2} \bullet 38$.

$$\frac{1}{2} \bullet (24 + 38) = \frac{1}{2} \bullet 24 + \frac{1}{2} \bullet 38$$

$$\frac{1}{2} \bullet 62 = 12 + 19$$

$$31 = 31$$

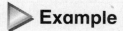 **Example**

What number makes $\left(4\frac{1}{3} \bullet 1\frac{3}{4}\right) \bullet 2\frac{1}{2} = 4\frac{1}{3} \bullet (\square) \bullet 2\frac{1}{2}$ a true statement?

Use the associative property of multiplication.

$$\left(4\frac{1}{3} \bullet 1\frac{3}{4}\right) \bullet 2\frac{1}{2} = 4\frac{1}{3} \bullet \left(\square \bullet 2\frac{1}{2}\right)$$

The number $1\frac{3}{4}$ makes $\left(4\frac{1}{3} \bullet 1\frac{3}{4}\right) \bullet 2\frac{1}{2} = 4\frac{1}{3} \bullet (\square) \bullet 2\frac{1}{2}$ a true statement.

You can use the properties of multiplication to help you find products.

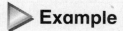 **Example**

What is $(-12 \bullet 1) \bullet (-8)$?

Use the multiplicative identity property of 1.

$$(-12 \bullet 1) \bullet (-8) = -12 \bullet (-8) = 96$$

$$(-12 \bullet 1) \bullet (-8) = 96$$

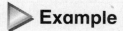 **Example**

What is $10 \bullet (5.9 + 4.7)$?

Use the distributive property of multiplication over addition to multiply each addend in the parentheses by 10.

$$10 \bullet (5.9 + 4.7) = 10 \bullet 5.9 + 10 \bullet 4.7$$

$$= 59 + 47 = 106$$

$$10 \bullet (5.9 + 4.7) = 106$$

TIP: The distributive property of multiplication over subtraction is true for rational numbers.

$$a \bullet (b - c) = a \bullet b - a \bullet c$$

Practice

Directions: For questions 1 through 6, write the property that is represented by the given equation.

1. $1 \cdot \frac{4}{5} = \frac{4}{5}$ _____

2. $3.4 \cdot 6.9 = 6.9 \cdot 3.4$ _____

3. $\left(2\frac{1}{2} \cdot 3\frac{3}{8}\right) \cdot 5\frac{1}{3} = 2\frac{1}{2} \cdot \left(3\frac{3}{8} \cdot 5\frac{1}{3}\right)$ _____

4. $-\frac{4}{5} \cdot \frac{3}{4} = \frac{3}{4} \cdot \left(-\frac{4}{5}\right)$ _____

5. $2.5 \cdot (2 \cdot 6.3) = (2.5 \cdot 2) \cdot 6.3$ _____

6. $\frac{1}{4} \cdot \left(\frac{5}{6} + \frac{1}{2}\right) = \frac{1}{4} \cdot \frac{5}{6} + \frac{1}{4} \cdot \frac{1}{2}$ _____

Directions: For questions 7 through 11, complete each equation to make it a true statement.

7. $(3.5 \cdot 6.4) \cdot 5.2 = 3.5 \cdot (6.4 \cdot$ _____$)$

8. $-17 \cdot$ _____ $= -17$

9. $1\frac{3}{10} \cdot 2\frac{4}{5} =$ _____ $\cdot 1\frac{3}{10}$

10. $\left(\frac{2}{3} \cdot \frac{1}{5}\right) \cdot 15 = \frac{2}{3} \cdot ($ _____ $\cdot 15)$

11. $2.8 \cdot (5.6 +$ _____ $) = 2.8 \cdot 5.6 + 2.8 \cdot 8.4$

CCSS: 7.NS.2.a, 7.NS.2.c

12. Which statement is true?

    A. $-8 \bullet (-12) = -12 \bullet 8$

    B. $\left(2\frac{2}{3} \bullet 4\frac{1}{2}\right) \bullet \frac{2}{9} = 2\frac{2}{3} \bullet \left(4\frac{1}{2} \bullet \frac{2}{9}\right)$

    C. $7.4 \bullet (2.5 + 3.9) = 7.4 \bullet 2.5 \bullet 2.9$

    D. $1 \bullet \left(-\frac{3}{4}\right) = \frac{3}{4} \bullet 0$

13. Use the associative property of multiplication to complete the equation. Then show that the two sides of the equation are equal.

    $(2.7 \bullet 4.1) \bullet 3.2 = $ _____

14. Use the distributive property of multiplication over addition to complete the equation. Then show that the two sides of the equation are equal.

    $\frac{5}{8} \bullet \left(\frac{4}{5} + \frac{3}{10}\right) = $ _____

15. Use the commutative property of multiplication to complete the equation. Then show that the two sides of the equation are equal.

    $1.5 \bullet 25.4 = $ _____

16. Fill in the missing information to find the product of $\frac{1}{3} \bullet \left(\frac{2}{3} \bullet \frac{9}{10}\right)$.

    $\frac{1}{3} \bullet \left(\frac{2}{3} \bullet \frac{9}{10}\right) = \left(\frac{1}{3} \bullet \frac{2}{3}\right) \bullet \frac{9}{10}$ by the _____ property of multiplication.

    $\left(\frac{1}{3} \bullet \frac{2}{3}\right) \bullet \frac{9}{10} = $ _____ $\bullet \frac{9}{10} = $ _____

17. Fill in the missing information to find the value of $2.6 \bullet (4.3 + 1)$.

    $2.6 \bullet (4.3 + 1) = 2.6 \bullet 4.3 + 2.6 \bullet 1$ by the _____ property of multiplication over addition.

    $2.6 \bullet 4.3 + 2.6 \bullet 1 = 2.6 \bullet 4.3 + 2.6$ by the _____ property of 1.

    $2.6 \bullet 4.3 + 2.6 = $ _____ $+ 2.6 = $ _____

CCSS: 7.NS.3

Lesson 6: Solving Problems with Rational Numbers

The following example shows the steps you can use to solve real-life problems involving rational numbers.

▶ **Example**

The temperature rose for 8 hours at a rate of 2°F per hour. What was the new temperature after 8 hours if the original temperature was −3°F?

Step 1: **Take time to study the problem.**
What does the problem want you to figure out?
You need to find the temperature after 8 hours.

Step 2: **Evaluate the information given.**
What information does the problem give?
The temperature rose for 8 hours at a rate of 2°F per hour.
The original temperature was −3°F.
Do you have all the information needed to solve the problem?
Yes, there is enough information to solve the problem.

Step 3: **Select the operation(s) needed to solve the problem.**
You need to use two operations:
multiplication (since the temperature rose the same amount for each of the 8 hours)
addition (to find the new temperature)

Step 4: **Do the math and check your answer.**

$8 \cdot 2 = 16$
$-3 + 16 = 13$

The answer is 13°F.

Check your answer by working backwards.

$13 - (8 \cdot 2) = 13 - 16 = -3$

The new temperature after 8 hours was 13°F.

CCSS: 7.NS.3

 # Example

The water level in a water tower fell at a rate of $1\frac{1}{2}$ feet per hour for $6\frac{1}{2}$ hours. What was the new water level if the original water level in the tower was 52 feet?

Step 1: **Take time to study the problem.**
What does the problem want you to figure out?
You need to find the water level after $6\frac{1}{2}$ hours.

Step 2: **Evaluate the information given.**
What information does the problem give?
The water level of a tower fell at a rate of $1\frac{1}{2}$ feet per hour for $6\frac{1}{2}$ hours.
The original water level in the tower was 52 feet.
Do you have all the information needed to solve the problem?
Yes, there is enough information to solve the problem.

Step 3: **Select the operation(s) needed to solve the problem.**
You need to use two operations:
multiplication (since the water level fell the same amount for each of the $6\frac{1}{2}$ hours)
subtraction (to find the new water level)

Step 4: **Do the math and check your answer.**

$$1\frac{1}{2} \cdot 6\frac{1}{2} = \frac{3}{2} \cdot \frac{13}{2} = \frac{39}{4} = 9\frac{3}{4}$$
$$52 - 9\frac{3}{4} = 51\frac{4}{4} - 9\frac{3}{4} = 42\frac{1}{4}$$

The answer is $42\frac{1}{4}$ feet.

Check your answer by working backwards.

$$42\frac{1}{4} + \left(1\frac{1}{2} \cdot 6\frac{1}{2}\right) = 42\frac{1}{4} + 9\frac{3}{4} = 52$$

The new water level at the end of $6\frac{1}{2}$ hours was $42\frac{1}{4}$ feet.

▶ Example

A box of donuts weighs 25.6 ounces. If the box weighs 4 ounces and each donut weighs 1.8 ounces, how many donuts are in the box?

Step 1: **Take time to study the problem.**
What does the problem want you to figure out?
You need to find the number of donuts in the box.

Step 2: **Evaluate the information given.**
What information does the problem give?
A box of donuts weighs 25.6 ounces.
The box weighs 4 ounces.
Each donut weighs 1.8 ounces.
Do you have all the information needed to solve the problem?
Yes, there is enough information to solve the problem.

Step 3: **Select the operation(s) needed to solve the problem.**
You need to use two operations:
subtraction (to subtract the weight of the box from the total weight)
division (to find the number of donuts in the box)

Step 4: **Do the math and check your answer.**
$25.6 - 4 = 21.6$
$21.6 \div 1.8 = 12$

The answer is 12 donuts.

Check your answer by working backwards.

$12 \cdot 1.8 + 4 = 21.6 + 4 = 25.6$

There are 12 donuts in the box.

CCSS: 7.NS.3

 Practice

Directions: For questions 1 through 6, write the operations needed to solve the problem. Then find the answer.

1. At the deli counter, Sandy bought $1\frac{3}{4}$ pounds of cheese for $4 per pound and $2\frac{1}{4}$ pounds of turkey for $6 per pound. How much did Sandy spend at the deli counter?

   operations: _____ and _____

   answer: _____

2. A carpenter cut pieces from a 12-foot-long board that were $4\frac{3}{4}$ feet long and $3\frac{1}{2}$ feet long. What is the length of the remaining part of the board?

   operations: _____ and _____

   answer: _____

3. Ann bought 2 packages of ground beef. Each package weighed $1\frac{3}{4}$ pounds. When she got home, she divided the ground beef into $\frac{1}{4}$-pound hamburger patties. How many hamburger patties did she make?

   operations: _____ and _____

   answer: _____

4. If 5 cans of soup cost $6.45, what is the cost of 8 cans of the soup?

   operations: _____ and _____

   answer: _____

5. At 6:00 P.M. the temperature was 9°F. The temperature dropped 2 degrees each hour for the next 6 hours. What was the temperature at midnight?

   operations: _____ and _____

   answer: _____

6. A ticket to an amusement park costs $9.25. Souvenir postcards cost $0.75 each. What is the total cost for a ticket to the amusement park and 12 postcards?

   operations: _____ and _____

   answer: _____

7. Gwen and Bob hiked along a 12-mile trail that had markers posted every $\frac{1}{8}$ mile. They hiked $4\frac{3}{8}$ miles and then stopped for lunch. How many markers will they pass after lunch when they finish the trail?

_____

8. A jeweler bought 8.5 ounces of gold at $1200 per ounce. The next day the value of gold increased to $1205 per ounce. What was the increase in the value of the jeweler's purchase?

_____

9. When Emily woke up, the temperature was 8°F. By noon, the temperature had risen by 3°F. It rose another 6°F in the afternoon to reach the day's high temperature. By the time Emily went to bed, the temperature had dropped 21°F from the high temperature. What was the temperature when Emily went to bed?

_____

10. In a football game, a team gained 8 yards on one play, then lost 4 yards, and then lost 3 yards on each of the next two plays. Which integer represents the total number of yards gained or lost on these plays?

    A. −18

    B. −2

    C. 2

    D. 18

11. A theater group sells tickets to a play. If an adult ticket costs $12.50 and a child ticket costs $8.75, then how much money will the group collect by selling 70 adult tickets and 15 child tickets?

    A. $1006.25

    B. $1000.00

    C. $875.00

    D. $131.25

12. Josh volunteers at a library. One weekend, he put 234 returned books back on the shelves. Of these books, $\frac{2}{3}$ were hardcover books and $\frac{1}{3}$ were paperback books.

   How many hardcover books did Josh put back on the shelves?

   _____

   How many more hardcover books than paperback books did Josh put back on the shelves?

   _____

   Explain how you found your answer.

   _____

   _____

   _____

13. A machinist needs to cut metal rods that are 3.75 inches long.

   How many rods of this length can the machinist cut from a 20-inch piece of metal?

   _____

   How much of the metal will be left over?

   _____

   Explain how you found your answer.

   _____

   _____

   _____

Unit 1 Practice Test

Directions: For questions 1 through 4, find the decimal expansion for the given number.

1. $\frac{-1}{2}$ _____

2. $\frac{1}{4}$ _____

3. $\frac{3}{8}$ _____

4. $\frac{5}{11}$ _____

Directions: For questions 5 through 8, use <, >, or = to compare the numbers.

5. 6.06 _____ $6\frac{3}{5}$

6. 7.44 _____ $7\frac{4}{9}$

7. 0.85 _____ $\frac{17}{20}$

8. $\frac{10}{3}$ _____ 3.34

9. Plot each point at its location on the number line. Write the number below each point.

$-1\frac{3}{5}$ 0.61 -2 $1\frac{1}{3}$

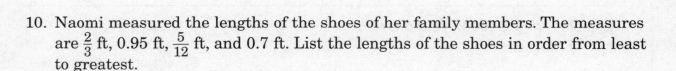

10. Naomi measured the lengths of the shoes of her family members. The measures are $\frac{2}{3}$ ft, 0.95 ft, $\frac{5}{12}$ ft, and 0.7 ft. List the lengths of the shoes in order from least to greatest.

_____

11. When Kareem went to France, he exchanged $225 for 175 euros. What decimal represents the number of euros he received for each dollar he exchanged? Round your answer to the nearest hundredth.

_____

Directions: For questions 12 and 13, use the number line to find the sum.

12. $0.5 + 0.8$

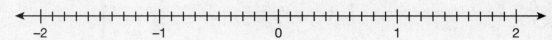

13. $\frac{-1}{8} + \left(-\frac{7}{8}\right)$

Directions: For questions 14 through 27, compute. Write answers in simplest form.

14. $-16 \bullet -7$ _____

15. $-0.8 + 3.2$ _____

16. $-\frac{7}{9} + \left(-\frac{1}{9}\right)$ _____

17. $\frac{-78}{6}$ _____

18. $5\frac{4}{7} + 2\frac{5}{7}$ _____

19. $\frac{7}{9} \div \frac{14}{15}$ _____

20. $1\frac{5}{7} \bullet \frac{7}{12}$ _____

21. $7.34 - 5.93$ _____

22. $\frac{5}{6} + \left(-\frac{5}{6}\right)$ _____

23. $49.77 \div 6.3$ _____

24. $12.3 \bullet 0.78$ _____

25. $\frac{9}{10} - \frac{3}{5}$ _____

26. $7\frac{5}{8} - 3\frac{3}{4}$ _____

27. $-63 \div -9$ _____

Directions: For questions 28 through 32, write the property that is represented by the given equation.

28. $\frac{2}{3} + 0 = \frac{2}{3}$ _____

29. $5.9 \bullet 3.1 = 3.1 \bullet 5.9$ _____

30. $2\frac{1}{2}\left(4\frac{5}{6} + 3\frac{3}{4}\right) = 2\frac{1}{2} \bullet 4\frac{5}{6} + 2\frac{1}{2} \bullet 3\frac{3}{4}$ _____

31. $\frac{-8}{9} \bullet 1 = \frac{-8}{9}$ _____

32. $\frac{5}{8} + \left(\frac{3}{8} + \frac{9}{10}\right) = \left(\frac{5}{8} + \frac{3}{8}\right) + \frac{9}{10}$ _____

Directions: For questions 33 through 36, complete each equation to make it a true statement.

33. $(2.5 \bullet 8.1) \bullet 6.3 = 2.5 \bullet (8.1 \bullet$ _____ $)$

34. _____ $+ (-12) = -12$

35. $1.6 \bullet (3.8 + 2.9) = 1.6 \bullet$ _____ $+ 1.6 \bullet 2.9$

36. $2\frac{1}{10} + 7\frac{3}{5} =$ _____ $+ 2\frac{1}{10}$

37. Fill in the missing information to find the sum of $\left(-1\frac{2}{3} + 2\frac{7}{10}\right) + 1\frac{2}{3}$.

$\left(-1\frac{2}{3} + 2\frac{7}{10}\right) + 1\frac{2}{3} = \left[2\frac{7}{10} + \left(-1\frac{2}{3}\right)\right] + 1\frac{2}{3}$ by the _____ property of addition.

$\left[2\frac{7}{10} + \left(-1\frac{2}{3}\right)\right] + 1\frac{2}{3} = 2\frac{7}{10} + \left(-1\frac{2}{3} + 1\frac{2}{3}\right)$ by the _____ property of addition.

$2\frac{7}{10} + \left(-1\frac{2}{3} + 1\frac{2}{3}\right) = 2\frac{7}{10} +$ _____

$2\frac{7}{10} +$ _____ $=$ _____ by the _____ property of 0.

38. The following sign shows the admission prices for the roller skating rink.

Roller Skating Admission

Adults	$5.00
Children (under 12)	$3.50

If a group of people spent $32.50 on tickets, which combination of tickets could they have bought?

A. 2 adults, 6 children

B. 3 adults, 5 children

C. 4 adults, 4 children

D. 5 adults, 3 children

39. A penguin dove to 130 feet below the water's surface. Then it swam up 48 feet toward the water's surface. Which integer represents where the penguin swam to?

A. 82 ft

B. −48 ft

C. −82 ft

D. −178 ft

40. What is $-9 \bullet 3 \bullet (-4)$?

A. −108

B. −10

C. 10

D. 108

41. Which statement below is always true?

A. The product of a number and zero is equal to that number.

B. The product of two negative numbers is negative.

C. The product of a positive number and a negative number is positive.

D. The product of two positive numbers is positive.

42. Look at the number sentence below.

$7.2 \bullet (9.1 + 4.3) =$
$7.2 \bullet 9.1 + 7.2 \bullet 4.3$

What number property does this number sentence represent?

A. distributive property of multiplication over addition

B. commutative property of multiplication

C. associative property of addition

D. associative property of multiplication

43. One lap around the park is 0.65 mile. Hal ran 4.5 laps. How many miles did Hal run?

A. 26.05

B. 4.95

C. 2.925

D. 2.605

44. Emily hiked along a trail that is 8 miles long. She hiked 2.25 miles before lunch, and then another 3.375 miles after lunch, before stopping for a rest. How many more miles did Emily need to hike to finish the trail?

_____

45. A baseball was pitched at a speed of 75.2 miles per hour. A tennis ball was served at a speed 1.5 times faster than this. How much faster was the speed of the tennis ball than the speed of the baseball?

_____

46. Ross used $1\frac{1}{2}$ gallons of paint on his kitchen walls. He used $1\frac{1}{3}$ times as much paint on his living room walls as he did on his kitchen walls. How many gallons of paint did Ross use in all to paint the kitchen and living room walls?

_____

47. Troy has $3\frac{1}{3}$ yards of computer cable. He uses $\frac{1}{6}$ yard of cable to connect his computer to other devices. How many connections can he make?

_____

48. Irene plays the viola. She practices $\frac{2}{3}$ hour every weekday and $1\frac{1}{4}$ hours each day on Saturday and Sunday. How many hours does she practice the viola each week?

_____

49. The temperature went from 10°F to −2°F in 4 hours. What was the average change in temperature per hour?

_____

52

50. Amy has 3 pieces of rope that have lengths of $15\frac{1}{3}$ feet, $20\frac{1}{2}$ feet, and $30\frac{3}{4}$ feet.

Part A
How many feet of rope does Amy have in all?

_____

Part B
If Amy cuts each length of rope into 3-foot pieces, how many 3-foot pieces of rope will she have?

_____

Part C
Explain how you found your answer.

_____

_____

_____

Part D
How much rope will be left over?

_____

Part E
Explain how you found your answer.

_____

_____

_____

51. A basket of apples weighs 78 ounces. If the basket weighs 12 ounces and each apple weighs 5.5 ounces, how many apples are in the basket?

Part A
Which operations will you use to solve the problem?

_____

Part B
Solve the problem and explain each step in your solution.

_____

_____

_____

_____

Part C
The basket can hold a total of 100 ounces. How many apples can the basket carry?

_____

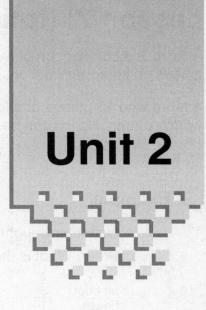

Unit 2

Ratios and Proportional Relationships

Do you have a favorite sports team? Then you may know that team rankings are determined by comparing the number of games won to the number of games played for each team. You can compare the win-loss ratios of two teams to see if they are equivalent. If they are equivalent ratios, they form a proportion.

In this unit, you will compute unit rates. You will decide whether two quantities are in a proportional relationship. You will identify the constant of proportionality. You will show proportional relationships between two quantities using tables, written descriptions, diagrams, graphs, and equations. Finally, you will use proportional relationships to solve ratio and percent problems.

In This Unit

Ratios and Rates

Proportional Relationships

Representing Proportional Relationships

Applications of Proportional Relationships

Lesson 7: Ratios and Rates

A **ratio** is a comparison of two numbers expressed as a fraction in simplest form. If a ratio is an improper fraction, it is left in this form and is **not** written as a mixed number.

A team won 13 games and lost 7 games. You can compare their wins and losses by writing a ratio. Ratios can be written in 3 ways.

13 to 7   13 : 7 $\frac{13}{7}$

▶ Example

A customer bought $\frac{2}{3}$ pound of roast beef and $\frac{1}{2}$ pound of cheese at the deli counter. What is the ratio of roast beef to cheese?

$$\frac{\text{roast beef}}{\text{cheese}} = \frac{\frac{2}{3}}{\frac{1}{2}}$$

To simplify the ratio, you can divide the numerator by the denominator.

$$\frac{2}{3} \div \frac{1}{2} = \frac{2}{3} \cdot \frac{2}{1} = \frac{4}{3}$$

The ratio of roast beef to cheese is 4 to 3.

▶ Example

A trail mix recipe calls for $\frac{1}{3}$ pound of mixed nuts, $\frac{4}{15}$ pound of raisins, and $\frac{2}{5}$ pound of granola. What is the ratio of raisins to mixed nuts?

Write a ratio.

$$\frac{\text{raisins}}{\text{mixed nuts}} = \frac{\frac{4}{15}}{\frac{1}{3}}$$

Divide the numerator by the denominator.

$$\frac{4}{15} \div \frac{1}{3} = \frac{4}{15} \cdot \frac{3}{1} = \frac{12}{15}$$

Write the fraction in simplest form.

$$\frac{12}{15} = \frac{12 \div 3}{15 \div 3} = \frac{4}{5}$$

The ratio of raisins to mixed nuts is 4 to 5.

CCSS: 7.RP.1

A **rate** is a ratio in which the terms have different units, such as 240 miles per 4 hours. A rate is called a **unit rate** when the second term is 1 unit, such as 60 miles per hour.

 Example

Riley can swim $\frac{1}{2}$ mile in $\frac{3}{4}$ hour. What is the unit rate of miles per hour?

Write a ratio.

miles per hour $= \dfrac{\frac{1}{2}}{\frac{3}{4}}$

Divide the numerator by the denominator.

$$\frac{1}{2} \div \frac{3}{4} = \frac{1}{2} \cdot \frac{4}{3} = \frac{4}{6} = \frac{2}{3}$$

The unit rate is $\frac{2}{3}$ mile per hour.

 Practice

Directions: Use the following information to answer questions 1 through 3.

The table shows how long Randi exercised each day.

Day	Monday	Tuesday	Wednesday	Thursday
Time Exercised	$\frac{2}{3}$ hour	$\frac{1}{2}$ hour	$\frac{1}{3}$ hour	$\frac{5}{6}$ hour

1. What is the ratio of the time Randi exercised on Monday to the time she exercised on Wednesday?

2. What is the ratio of the time Randi exercised on Tuesday to the time she exercised on Thursday?

3. What is the ratio of the time Randi exercised on Monday to the total time she exercised on all four days?

4. Which of the following has a ratio of $\frac{3}{4}$?

   A. $\frac{1}{2}$ to $\frac{2}{3}$

   B. $\frac{1}{4}$ to $\frac{2}{3}$

   C. $\frac{1}{3}$ to $\frac{1}{4}$

   D. $\frac{2}{3}$ to $\frac{1}{4}$

5. An artist made purple paint by mixing $\frac{1}{2}$ quart of red paint and $\frac{3}{4}$ quart of blue paint. What is the ratio of red paint to blue paint?

   A. $\frac{1}{3}$ quart of red to 1 quart of blue

   B. $\frac{2}{3}$ quart of red to 1 quart of blue

   C. $\frac{3}{4}$ quart of red to 1 quart of blue

   D. $\frac{4}{3}$ quart of red to 1 quart of blue

6. Melba rode her bike $\frac{7}{8}$ mile in $\frac{1}{4}$ hour. What is the unit rate in miles per hour?

   _____

7. It took a farmer $7\frac{1}{2}$ hours to plant $\frac{1}{8}$ acre with plants. What is the unit rate in acres per hour?

   _____

8. George walked $\frac{1}{2}$ mile in $8\frac{1}{2}$ minutes. What is the unit rate in miles per minute?

   _____

9. A carpet layer can put down $\frac{7}{8}$ square yard of carpeting in $\frac{1}{20}$ hour. What is the unit rate in square yards per hour?

   _____

10. A baker can prepare $2\frac{1}{2}$ dozen muffins in $\frac{1}{4}$ hour.

    Write a unit rate for this. _____

    Explain how you found your answer.

    _____

    _____

CCSS: 7.RP.2.a, 7.RP.2.b, 7.RP.2.c

Lesson 8: Proportional Relationships

A **proportion** states that two ratios are equal. You can verify whether two ratios demonstrate a proportional relationship.

 Example

Tell whether $\frac{4}{16}$ and $\frac{5}{20}$ demonstrate a proportional relationship.

Write each ratio in simplest form and compare.

$$\frac{4}{16} = \frac{1}{4}$$

$$\frac{5}{20} = \frac{1}{4}$$

Since both $\frac{4}{16}$ and $\frac{5}{20}$ are equal to $\frac{1}{4}$, they are equal ratios and form a proportional relationship.

A **proportion** has **cross products** that are equal to each other.

$$\frac{4}{16} = \frac{5}{20}$$

$$4 \bullet 20 = 16 \bullet 5$$

$$80 = 80$$

$\frac{4}{16} = \frac{5}{20}$ because the cross products are equal.

▶ **Example**

Tell whether $\frac{5}{6}$ and $\frac{12}{15}$ demonstrate a proportional relationship.

Find the cross products.

$$\frac{5}{6} \overset{?}{=} \frac{12}{15}$$

$$5 \bullet 15 \overset{?}{=} 6 \bullet 12$$

$$75 \neq 72$$

Since $\frac{5}{6}$ and $\frac{12}{15}$ do not have equal cross products, they do not form a proportional relationship.

You can use cross products to find missing values in proportions. This method of solving a proportion is called **cross multiplication**.

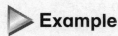 **Example**

What is the missing value in the following proportion?

$$\frac{16}{20} = \frac{n}{35}$$

$$\frac{16}{20} = \frac{n}{35}$$

$$16 \cdot 35 = n \cdot 20$$

$$560 = 20n$$

$$28 = n$$

The missing value is 28.

Two quantities are directly proportional if their values increase or decrease together at the same ratio.

For example, suppose T-shirts are on sale for $5 each. Four T-shirts would cost 4 • $5, or $20. Eight T-shirts would cost 8 • $5, or $40. Doubling the number of T-shirts doubles the total cost. Changing the number of T-shirts changes the total cost. The equation $c = n \cdot 5$, or $c = 5n$, where c represents the total cost and n is the number of T-shirts, is an equation which represents this proportional relationship. The value of the constant, in this case the cost of a T-shirt or $5, is called the **constant of proportionality.**

Example

How much would it cost to buy 10 T-shirts?

Use the proportional equation $c = 5n$.
$$c = 5n$$
$$= 5 \cdot 10$$
$$= 50$$

It would cost $50 to buy 10 T-shirts.

CCSS: 7.RP.2.a, 7.RP.2.b, 7.RP.2.c

The distance formula $d = rt$, where d represents distance, r represents the speed, and t represents the time, is a proportional equation.

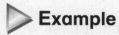 **Example**

If Zachary drives at a constant speed of 55 miles per hour, how many hours would it take him to drive 165 miles?

Use the distance formula $d = rt$.
Substitute the known values for the variables.

$$d = rt$$
$$165 = 55t$$
$$3 = t$$

It would take Zachary 3 hours to drive 165 miles.

A **unit price** is a price for one unit of a product. The unit price of an item can be its price per ounce, per pound, per hour, or any unit of measure. Some unit prices are $0.19 per ounce, $3.99 per pound, and $8.00 per hour. Notice that these unit prices could also be written as $0.19 per 1 ounce, $3.99 per 1 pound, and $8.00 per 1 hour.

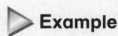 **Example**

Mrs. Lippman bought 16 gallons of gasoline for $52. What is the price per gallon of the gasoline she bought?

Use the ratio of dollars to gallons of gas to find the unit price.

$$\frac{\text{dollars}}{\text{gallons of gas}} = \frac{52}{16} = 3.25$$

Mrs. Lippman paid $3.25 per gallon of gas.

► Example

Julie works as a bagger in a grocery store. Last week she worked for 20 hours and was paid $150. How much does Julie earn per hour?

Use the ratio of dollars to hours to find the unit price.

$$\frac{\text{dollars}}{\text{hours}} = \frac{150}{20} = 7.50$$

Julie earns $7.50 per hour.

CCSS: 7.RP.2.a, 7.RP.2.b, 7.RP.2.c

You can compare the unit prices of similar products to find the better deal.

▶ Example

Crunch's Peanut Butter costs $2.66 for 14 ounces. Nutty's Peanut Butter costs $3.36 for 16 ounces. Which brand of peanut butter is the better deal?

Find the unit price of Crunch's Peanut Butter.

$$\frac{\text{dollars}}{\text{ounces}} = \frac{2.66}{14} = 0.19$$

Crunch's Peanut Butter costs $0.19 for 1 ounce.

Find the unit price of Nutty's Peanut Butter.

$$\frac{\text{dollars}}{\text{ounces}} = \frac{3.36}{16} = 0.21$$

Nutty's Peanut Butter costs $0.21 for 1 ounce.

Therefore, Crunch's Peanut Butter is the better deal because it costs less for each ounce of peanut butter.

▶ Example

Sturdy's paper plates cost $3.12 for 24 plates. Paperoni's paper plates cost $4.32 for 36 paper plates. Which brand of paper plates is the better deal?

Find the unit price of Sturdy's paper plates.

$$\frac{\text{dollars}}{\text{plates}} = \frac{3.12}{24} = 0.13$$

Sturdy's paper plates cost $0.13 per plate.

Find the unit price of Paperoni's paper plates.

$$\frac{\text{dollars}}{\text{plates}} = \frac{4.32}{36} = 0.12$$

Paperoni's paper plates cost $0.12 per plate.

Therefore, Paperoni's paper plates are the better deal because they cost less per plate.

CCSS: 7.RP.2.a, 7.RP.2.b, 7.RP.2.c

⬤ Practice

Directions: For questions 1 through 8, write whether the ratios demonstrate a proportional relationship.

1. $\frac{4}{7}$ and $\frac{20}{28}$ _____

2. $\frac{3}{9}$ and $\frac{6}{18}$ _____

3. $\frac{15}{9}$ and $\frac{20}{12}$ _____

4. $\frac{12}{15}$ and $\frac{6}{10}$ _____

5. $\frac{10}{30}$ and $\frac{3}{12}$ _____

6. $\frac{20}{12}$ and $\frac{15}{9}$ _____

7. $\frac{24}{9}$ and $\frac{15}{6}$ _____

8. $\frac{8}{18}$ and $\frac{20}{45}$ _____

9. Carlos walked 8 miles in 3 hours. Leah walked 14 miles in 6 hours. Are these rates in proportion?

_____

10. Mark has two differently-sized prints of the same photograph. One print is 8 in. by 10 in. The other print is 11 in. by 14 in. Are these two prints in proportion?

_____

11. In a basketball game, Samantha made 15 shots out of 25 attempts. Parul made 12 shots out of 20 attempts. Are these rates in proportion?

_____

12. A middle-school basketball team won 8 of its first 12 games. The team later won 14 of its last 21 games. Do these ratios form a proportion?

_____

Directions: For questions 13 through 16, find the missing value in each proportion.

13. $\frac{2}{3} = \frac{n}{27}$ _____

14. $\frac{6}{3} = \frac{20}{n}$ _____

15. $\frac{n}{36} = \frac{8}{24}$ _____

16. $\frac{3}{n} = \frac{15}{25}$ _____

Directions: For questions 17 through 20, use the equation $c = 6n$, where c represents the cost of n items that sell for $6 each.

17. What is the constant of proportionality? _____

18. What will be the cost of 2 items? _____

19. What will be the cost of 6 items? _____

20. Explain why the cost of 6 items is proportional to the cost of 2 items.

_____

_____

21. Raffle tickets cost $2 each. Write an equation that shows the total cost, c, of buying r raffle tickets.

_____

22. The speed limit on a highway is 65 miles per hour. Write an equation that shows the total number of miles driven, d, in t hours.

_____

23. Mr. Strand's car gets 23 miles per gallon. Write an equation that shows the total number of miles, m, Mr. Strand can drive on g gallons of gas.

_____

24. Keiko earns $9 per hour at her job. Write an equation that shows how much she earns, e, if she works w hours.

_____

Directions: For questions 25 through 32, find the unit price.

25. Andrew bought 2 shirts for $31.00. _____

26. Georgia bought a 5-pound bag of apples for $4.00. _____

27. Mrs. Kelly purchased 6 spools of thread for $22.50. _____

28. Jane paid $26.00 for 4 dance classes. _____

29. Keisha earned $154 for 14 hours of work. _____

30. Henry paid $25.98 for 2 CDs. _____

31. Ari bought a 16-ounce bottle of iced tea for $1.28. _____

32. Martin paid $10.80 for a bouquet of a dozen flowers. _____

Directions: For questions 33 through 36, compare the two unit prices to find the better deal.

33. Six muffins cost $7.50. Eight muffins cost $9.60. Which is the better deal?

_____

34. An 18-ounce box of raisins costs $2.88. A 24-ounce box of raisins costs $3.60. Which is the better deal?

_____

35. Three yards of fabric costs $7.47. Seven yards of fabric costs $18.13. Which is the better deal?

_____

36. A 4-pack of pens costs $6.28. A 3-pack of pens costs $4.65. Which is the better deal?

_____

37. Bags of flour come in different sizes. Which is the best buy?

   A. a 2-lb bag of flour for $0.88

   B. a 3-lb bag of flour for $1.35

   C. a 5-lb bag of flour for $2.30

   D. a 10-lb bag of flour for $4.10

38. Jars of spaghetti sauce come in different sizes. Which is the best buy?

   A. a 26-oz jar of spaghetti sauce for $2.86

   B. an 18-oz jar of spaghetti sauce for $2.16

   C. a 48-oz jar of spaghetti sauce for $4.80

   D. a 32-oz jar of spaghetti sauce for $4.16

39. Bobby can buy a 15-ounce bottle of shampoo for $2.85 or a 20-ounce bottle of the same shampoo for $3.80. He can use a $0.60-off coupon for either bottle of shampoo.

   What is the price per ounce of the 15-ounce bottle of shampoo with the coupon?

   _____

   Explain how you found your answer.

   _____

   _____

   _____

   What is the price per ounce of the 20-ounce bottle of shampoo with the coupon?

   _____

   Explain how you found your answer.

   _____

   _____

   _____

   Which is the better buy?

   _____

CCSS: 7.RP.2.a, 7.RP.2.b, 7.RP.2.d

Lesson 9: Representing Proportional Relationships

There are many ways to show that a proportional relationship exists between two quantities. You have already learned how cross products can be used to demonstrate a proportional relationship between two ratios. You can also use tables, written descriptions, diagrams, and graphs to demonstrate proportional relationships.

 Example

The table gives pairs of values for the variables x and y.

x	1	2	3
y	3	6	9

Determine whether the values in the table form a proportional relationship.

Find the ratio of each y-value to its corresponding x-value.

$$\frac{3}{1} \qquad \frac{6}{2} = \frac{3}{1} \qquad \frac{9}{3} = \frac{3}{1}$$

Since all of the given pairs of values form the constant ratio of $\frac{3}{1}$, the values demonstrate a proportional relationship.

 Example

The table gives pairs of values for the variables a and b, which are in a proportional relationship.

a	1	2	3
b	5	10	15

Find the constant of proportionality and write an equation that represents the table.

Compare each b-value to its corresponding a-value.

$$\frac{5}{1} \qquad \frac{10}{2} = \frac{5}{1} \qquad \frac{15}{3} = \frac{5}{1}$$

The given pairs of values result in a constant ratio of $\frac{5}{1}$ or 5, which is the constant of proportionality.

Since the ratio $\frac{b}{a}$ is always the same, $\frac{5}{1}$, $\frac{b}{a} = \frac{5}{1}$. The proportion $\frac{b}{a} = \frac{5}{1}$ can be written as $b = 5a$ by finding the cross products.

When given a verbal description of a proportional relationship, you can find the constant of proportionality by finding the unit rate.

 Example

Gina can type 100 words in 2 minutes. What is the constant of proportionality?

Compare the number of words to the number of minutes.

$$\frac{100 \text{ words}}{2 \text{ minutes}} = \frac{50 \text{ words}}{\text{minute}}$$

The unit rate is $\frac{50 \text{ words}}{\text{minute}}$. The relationship between the number of words typed, n, and the number of minutes, t, is $n = 50t$.

The constant of proportionality is 50.

You can determine whether a diagram shows a proportional relationship.

 Example

Determine whether the lengths of the corresponding sides of the rectangles below form a proportional relationship.

9 in.   I   6 in.   II

12 in.   8 in.

Compare the lengths of the rectangles.

$$\frac{\text{length of rectangle I}}{\text{length of rectangle II}} = \frac{12}{8} = \frac{3}{2}$$

Compare the widths of the rectangles.

$$\frac{\text{width of rectangle I}}{\text{width of rectangle II}} = \frac{9}{6} = \frac{3}{2}$$

Since both ratios are equal, the rectangles demonstrate a proportional relationship.

 TIP: If all of the corresponding sides of two figures form a proportional relationship, the figures are called similar figures. Similar figures have the same shape but different sizes.

68

CCSS: 7.RP.2.a, 7.RP.2.b, 7.RP.2.d

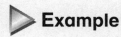 **Example**

Examine these similar triangles, and find the constant of proportionality.

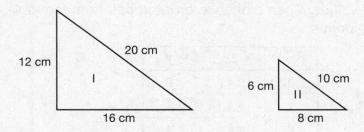

Compare the lengths of the corresponding sides.

$$\frac{\text{shortest side of } \triangle\text{II}}{\text{shortest side of } \triangle\text{I}} = \frac{6}{12} = \frac{1}{2}$$

$$\frac{\text{middle side of } \triangle\text{II}}{\text{middle side of } \triangle\text{I}} = \frac{8}{16} = \frac{1}{2}$$

$$\frac{\text{longest side of } \triangle\text{II}}{\text{longest side of } \triangle\text{I}} = \frac{10}{20} = \frac{1}{2}$$

The lengths of the corresponding sides have a ratio of $\frac{1}{2}$.

The constant of proportionality is $\frac{1}{2}$.

An equation for a proportional relationship can be represented by a line on a coordinate plane. The line passes through the point (0, 0) and the ratio of the **y-coordinate** to the **x-coordinate** for every other point on the graph is equal to the constant of proportionality. On a line that passes through (0, 0), the *y*-coordinate of a point whose *x*-coordinate is 1 is equal to the **unit rate**.

 TIP: In a coordinate plane, the point (0, 0) is often referred to as the **origin**.

Example

Graph $y = 2x$ on a coordinate plane. Identify the unit rate from the graph.

Make a table of values. Then plot each ordered pair from the table on a graph and connect the points.

x	2x	y
0	2 • 0	0
1	2 • 1	2
2	2 • 2	4
3	2 • 3	6
4	2 • 4	8

The ordered pairs are (0, 0), (1, 2), (2, 4), (3, 6), and (4, 8).

The line passes through (0, 0), so the y-coordinate of the ordered pair whose x-coordinate is 1 is equal to the unit rate. On the graph to the right, the point (1, 2) shows that 2 is the unit rate.

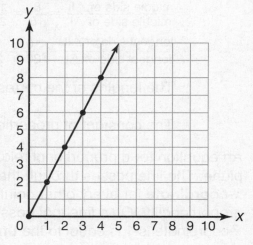

Notice that the ratio of the y-coordinate to the x-coordinate for each point (other than the origin) on the graph to the right is equal to $\frac{2}{1}$ or 2, the unit rate.

 Example

Determine if the graph below shows a proportional relationship.

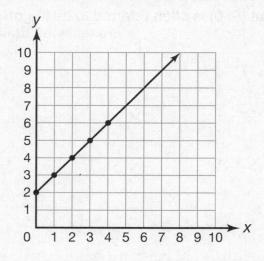

The graph shows a straight line, but it does not pass through (0, 0). Therefore, the graph does not show a proportional relationship.

CCSS: 7.RP.2.a, 7.RP.2.b, 7.RP.2.d

 Practice

Directions: For questions 1 through 4, determine whether the values in the table form a proportional relationship.

1.

a	12	8	4
b	3	2	1

_____

2.

p	4	5	6
q	6	7	8

_____

3.

r	1	2	3
s	1	4	9

_____

4.

g	2	4	6
h	3	6	9

_____

Directions: For questions 5 and 6, write an equation that represents the table.

5.

x	9	6	3
y	3	2	1

_____

6.

x	1	2	3
y	4	8	12

_____

Directions: For questions 7 and 8, find the constant of proportionality.

7. A bicycle rental at the beach costs $36 for 3 days. _____

8. An engine is turning at the rate of 2800 revolutions in 4 minutes. _____

Directions: For questions 9 through 11, determine whether the lengths of the corresponding sides of the figures form a proportional relationship.

9.

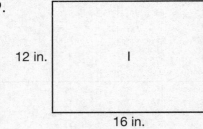 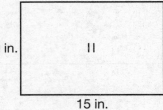

_____

10.

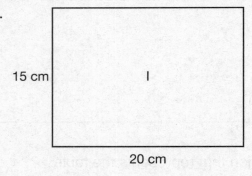 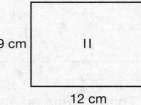

_____

11.

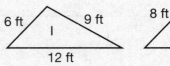

_____

CCSS: 7.RP.2.a, 7.RP.2.b, 7.RP.2.d

Directions: For questions 12 and 13, graph the equation on a coordinate plane. Identify the unit rate from the graph.

12. $y = 3x$

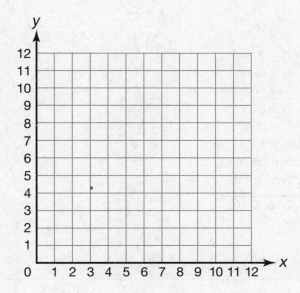

unit rate: _____

13. $y = 4x$

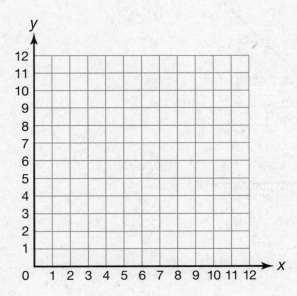

unit rate: _____

CCSS: 7.RP.2.a, 7.RP.2.b, 7.RP.2.d

Directions: For questions 14 and 15, determine if the graph shows a proportional relationship.

14.

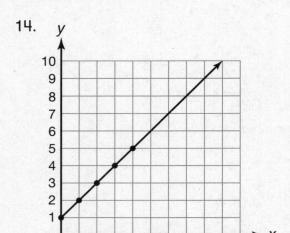

_____

15.

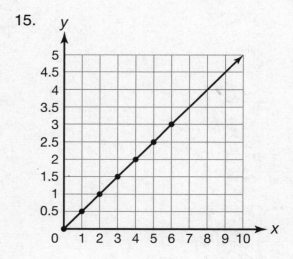

_____

CCSS: 7.RP.2.a, 7.RP.2.b, 7.RP.2.d

16. For which statement is the constant of proportionality equal to 8?

    A. Lisa lost 8 pounds in 8 weeks.

    B. Frank drove 400 miles in 8 hours.

    C. There are 80 people in 10 rows.

    D. There are 100 boxes on 8 shelves.

17. What is the equation for the table below?

x	0	1	2	3	4
y	0	7	14	21	28

    A. $y = x + 7$

    B. $y = 7x$

    C. $x = 7y$

    D. $y = \frac{1}{7}x$

18. Use the table below to answer the questions.

x	0	1	2	3	4	5
y	0	1	2	3	4	5

Plot the points on a coordinate grid and connect the points.

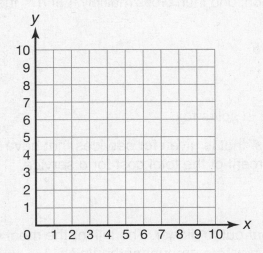

What is the unit rate?

_____

Explain how you found your answer.

_____

_____

Lesson 10: Applications of Proportional Relationships

In this lesson you will use proportions to find taxes and gratuities, markups and markdowns on items for sale, commissions and fees, simple interest, percent increase or decrease, and percent error.

Proportions can be used to solve problems involving percents. The following proportion can be used to find an unknown value.

$$\frac{part}{whole} = \frac{percent}{100}$$

Sales Tax and Gratuities

Percents are often used when describing and solving problems involving sales tax and gratuities. The sales tax rate varies by state, and sometimes, by city. It is a percent of the total cost of items purchased.

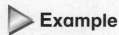 **Example**

> Larry bought a video game console for $129. The sales tax in his state is 6%. How much tax did Larry pay for the video game console?
>
> Set up a proportion, and then cross multiply. Let $n =$ the amount of the sales tax.
>
> $$\frac{n}{129} = \frac{6}{100}$$
> $$100 \cdot n = 129 \cdot 6$$
> $$100n = 774$$
> $$n = 7.74$$
>
> Larry paid $7.74 in sales tax.

A gratuity is a tip, or bonus, that is given for services that have been provided. It is usually computed as a percent of the total cost for a service.

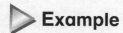 **Example**

> Kyle's family went out to dinner. The bill for the meal was $45. Kyle's dad left a 15% tip. What was the amount of the tip?
>
> Set up a proportion, and then cross multiply. Let $n =$ the amount of the tip.
>
> $$\frac{n}{45} = \frac{15}{100}$$
> $$100 \cdot n = 45 \cdot 15$$
> $$100n = 675$$
> $$n = 6.75$$
>
> The amount of the tip was $6.75.

CCSS: 7.RP.3

Markups and Markdowns

A **markup** is the difference between the cost of an item and its selling price.
A **markdown** is the amount an item is reduced from its regular price.

If you know the cost of an item and the markup, you can find the selling price.
selling price = cost + markup

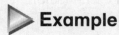 **Example**

A store buys sweatshirts for $12 each and marks up the price by 25%. What is the price for a sweatshirt at this store?

Find the markup. Set up a proportion, and then cross multiply.
Let n = the amount of the markup.

$$\frac{n}{12} = \frac{25}{100}$$

$100 \cdot n = 12 \cdot 25$

$100n = 300$

$n = 3$

The markup is $3.

Add the markup to the cost.
$12 + $3 = $15

The price for a sweatshirt at this store is $15.

If you know the original price of an item and the markdown, you can find the sale price.
sale price = original price − markdown

Example

The original price of a pair of sneakers was $49. The price of the sneakers is marked down 30%. What is the sale price of the sneakers?

Find the markdown. Set up a proportion, and then cross multiply.
Let n = the amount of the markdown.

$$\frac{n}{49} = \frac{30}{100}$$

$100n = 1470$

$n = 14.70$ The markdown is $14.70.

Subtract the markdown from the original price.
$49.00 − $14.70 = $34.30

The sale price of the sneakers is $34.30.

Commissions and Fees

A **commission** is an amount of money paid to a salesperson based on his or her total amount of sales and the commission rate.

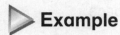 **Example**

Richard is paid 8% commission on his carpet sales. How much commission will Richard earn for sales of $1650?

Set up a proportion, and then cross multiply. Let n = the amount of the commission.

$$\frac{n}{1650} = \frac{8}{100}$$
$$100 \cdot n = 1650 \cdot 8$$
$$100n = 13{,}200$$
$$n = 132$$

Richard will earn $132 in commission.

A **fee** is an amount added on to the cost of an item for a service.

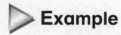 **Example**

Mindy bought a birthday present for a friend for $15 online. This included a shipping and handling fee of $3. What percent of the cost of the present was the shipping and handling fee?

Set up a proportion, and then cross multiply. Let n = the percent.

$$\frac{3}{15} = \frac{n}{100}$$
$$3 \cdot 100 = 15 \cdot n$$
$$300 = 15n$$
$$20 = n$$

$$\frac{20}{100} = 20\%$$

The shipping and handling fee was 20% of the cost of the present.

CCSS: 7.RP.3

Interest Rate and Simple Interest

Interest is the amount you are charged when you borrow money or the amount you are paid when you invest money. The **principal** is the original amount invested or borrowed. The **interest rate** is a percent of the principal. **Simple interest** is paid only on the principal and is paid at the end of an investment time period.

To find simple interest, use the formula $I = prt$ where $I =$ the amount of interest, $p =$ the principal, $r =$ the interest rate, and $t =$ the time in years.

 Example

Tom borrowed $300 to purchase a mountain bike. If he pays interest at a rate of 4% on a 6-month loan, how much interest will he pay?

$I = prt$ Use the simple interest formula.

$= 300 \cdot 4\% \cdot \frac{1}{2}$ $p = 300$, $r = 4\%$, $t = 6$ mo $= \frac{1}{2}$ yr

$= 300 \cdot 0.04 \cdot 0.5$

$= 6$

Tom will pay $6 in interest on his loan.

 Example

Victoria deposited $2500 into a savings account that pays her simple interest. The interest rate is 3%. How much interest will Victoria earn in 2 years? What will Victoria's account balance be after 2 years?

$I = prt$ Write the rate as a decimal and multiply.

$= 2500 \cdot 3\% \cdot 2$

$= 2500 \cdot 0.03 \cdot 2$ $p = 2500$, $r = 3\% = 0.03$, $t = 2$ yr

$= 150$

Then, find the account balance by adding the principal and the interest.
$2500 + 150 = 2650$

After 2 years, Victoria will earn $150 interest. After two years, her account balance will be $2650.

Percent of Increase or Decrease

The **percent of increase** or **percent of decrease** is the change from a given amount expressed as a percent of that amount.

To find the percent of increase, use the following proportion:

$$\frac{\text{new amount} - \text{original amount}}{\text{original amount}} = \frac{n}{100}$$

 Example

In one year, Braden's height went from 60 inches to 63 inches. What was the percent of increase in Braden's height?

Set up a proportion, and then cross multiply. Let n = the percent of increase.

$$\frac{\text{new amount} - \text{original amount}}{\text{original amount}} = \frac{n}{100}$$

$$\frac{63 - 60}{60} = \frac{n}{100}$$

$$\frac{3}{60} = \frac{n}{100}$$

$$3 \cdot 100 = 60 \cdot n$$

$$300 = 60n$$

$$5 = n$$

$$\frac{5}{100} = 5\% \qquad \text{Braden's height increased by 5\% in one year.}$$

To find the percent of decrease, use the following proportion:

$$\frac{\text{original amount} - \text{new amount}}{\text{original amount}} = \frac{n}{100}$$

 Example

Tara purchased a DVD player for $150. Six months later, the same DVD player was selling for $120. What was the percent of decrease in the price?

Set up a proportion, and then cross multiply. Let n = the percent of decrease.

$$\frac{\text{original amount} - \text{new amount}}{\text{original amount}} = \frac{n}{100}$$

$$\frac{150 - 120}{150} = \frac{n}{100}$$

$$\frac{30}{150} = \frac{n}{100}$$

$$30 \cdot 100 = 150 \cdot n$$

$$3000 = 150n$$

$$20 = n$$

$$\frac{20}{100} = 20\% \qquad \text{The price of the DVD player decreased by 20\%.}$$

CCSS: 7.RP.3

Percent Error

Percent error describes how close a **measured** or **estimated value** is to the **actual value**. To find the percent error, use the following proportion:

$$\frac{|\text{measured or estimated value} - \text{actual value}|}{\text{actual value}} = \frac{n}{100}$$

 Example

George estimated that there were 315 students at a basketball game. The actual number of students at the game was 350. What was the percent error of his estimate?

Set up a proportion, and then cross multiply. Let $n =$ the percent error.

$$\frac{|\text{estimated value} - \text{actual value}|}{\text{actual value}} = \frac{n}{100}$$

$$\frac{|315 - 350|}{350} = \frac{n}{100}$$

$$\frac{35}{350} = \frac{n}{100}$$

$$35 \cdot 100 = 350 \cdot n$$

$$3500 = 350n$$

$$10 = n$$

$$\frac{10}{100} = 10\%$$

The percent error of George's estimate was 10%.

Example

In an experiment, a chemist estimated the amount of liquid in a beaker to be 52 milliliters. The actual amount of liquid in the beaker was 50 milliliters. What was the percent error?

Set up a proportion, and then cross multiply. Let $n =$ the percent error.

$$\frac{|\text{estimated value} - \text{actual value}|}{\text{actual value}} = \frac{n}{100}$$

$$\frac{|52 - 50|}{50} = \frac{n}{100}$$

$$\frac{2}{50} = \frac{n}{100}$$

$$2 \cdot 100 = 50 \cdot n$$

$$200 = 50n$$

$$4 = n$$

$$\frac{4}{100} = 4\%$$

The percent error was 4%.

Practice

1. Jesse is buying a stereo that costs $799. If the sales tax is 7%, how much tax will Jesse have to pay for the stereo?

   _____

2. Heather bought a CD on sale for $11 and a book that cost $14. The sales tax rate is 5%. How much tax did Heather pay for both items?

   _____

3. Juan bought a laptop for $650 and a power adapter for $59. If the sales tax rate is 6%, what was the total tax that Juan paid?

   _____

4. Patricia took her family out to breakfast. The total cost of the food was $24. She left a 15% tip for the waiter. How much did Patricia leave for a tip?

   _____

5. Logan and his father each have a sandwich and a drink for lunch at a restaurant. A sandwich costs $5.25 and a drink costs $1.75. Logan's father leaves a 20% tip. What is the amount of the tip?

   _____

6. The regular price of a DVD is $24. During a sale, the price of a DVD is marked down 30%. What will be the sale price of the DVD?

   _____

7. A toy store marks up its prices by 45%. What will be the price of a doll that the toy store bought for $12?

_____

8. At a clearance sale, Jolissa bought a jacket that was marked down 60%. The regular price of the jacket was $89. How much did Jolissa pay for the jacket?

_____

9. Theo owns a hardware store. He marks up the price of all tools by 35%. What would be the price of a hammer at Theo's store if it cost him $30?

_____

10. A clothing store is having an anniversary sale. Every piece of clothing in the store is marked down 15%. What is the sale price of a belt that has a regular price of $23?

_____

11. Sara earns a 15% commission on furniture that she sells. How much commission would she earn if she sells a dining room set for $850?

_____

12. Dmitri is a real estate agent. He earns a 3% commission on each house he sells. How much commission will Dmitri earn if he sells a house for $270,000?

_____

13. Joan receives a 9% commission on appliances she sells. Last month she sold $18,500 worth of appliances. How much did Joan earn in commissions last month?

_____

14. Eddie sold an antique lamp on an online auction website for $75. Eddie had to pay the website a fee of $6 on the final sale price. What percent of the sale price was the fee to the auction website?

_____

15. Monica sent $250 to her family overseas using an Internet service. She paid an additional fee of $15 to send the money electronically. What percent of the money sent was the fee to the Internet service?

_____

16. Tamiko earns 3% simple interest on the balance of her savings account. Her account had a balance of $3500. How much simple interest will Tamiko earn on the balance after 6 months?

_____

17. Pedro took out a 4-year loan for $32,500 to help pay his college tuition. He has to pay 4% simple interest on the loan. How much interest will Pedro have to pay?

_____

18. Jin took out a 6-year loan for $18,000 to buy a new car. He has to pay 2% simple interest on the loan. How much will Jin have paid in all after the 6 years?

_____

CCSS: 7.RP.3

19. Last year there were 520 students enrolled at Olsen Middle School. This year there are 572 students enrolled. What is the percent of increase in the number of students?

_____

20. Matt's savings account balance went from $325 at the beginning of the month to $195 at the end of the month. What was the percent of decrease?

_____

21. The value of a car that cost $16,000 decreased to $12,000 after two years. What was the percent of decrease?

_____

22. Ana estimated the mass of an object to be 54 grams. The actual mass of the object was 45 grams. What was the percent error of Ana's estimate?

_____

23. Dane estimated that there were 420 people at a concert with him. The actual number of people at the concert was 375. What was the percent error of Dane's estimate?

_____

24. Nancy mistakenly measured the temperature of a heated liquid to be 48°C. The actual temperature of the heated liquid was 50°C. What was the percent error?

_____

25. Mrs. Piazza invested $8000 in a certificate of deposit that pays 3% simple interest for 4 years. How much will be in Mrs. Piazza's account at the end of 4 years?

    A. $960

    B. $7040

    C. $8240

    D. $8960

26. Last year a real estate broker sold 90 homes. This year she sold 63 homes. What was the percent of decrease in the number of homes she sold?

    A. 27%

    B. 30%

    C. 63%

    D. 70%

27. A computer normally sells for $520.

    During a sale, the computer is marked down 15%. How much will it cost to buy the computer during the sale?

    _____

    Explain how you found your answer.

    _____

    _____

    The sales tax rate is 6%. How much tax will be paid on the computer during the sale?

    _____

    Explain how you found your answer.

    _____

    _____

Unit 2 Practice Test

1. Ben is buying a refrigerator that costs $849. If the sales tax rate is 8%, how much tax will Ben have to pay on the refrigerator?

   _____

2. The table shows how long it took Melinda to do her math homework each day over four days.

Day	Monday	Tuesday	Wednesday	Thursday
Time Spent on Homework (in hours)	$\frac{3}{4}$	$\frac{5}{6}$	$\frac{2}{3}$	$\frac{1}{2}$

   What is the ratio of the time it took Melinda to do her math homework on Wednesday to the time it took her to do it on Tuesday?

   _____

3. The regular price of a sweater is $32. For a sale, the price of the sweater is marked down 30%. What is the sale price of the sweater?

   _____

4. Cara hiked 14 miles in 4 hours. Leslie hiked 21 miles in 6 hours. Do these rates demonstrate a proportional relationship?

   _____

5. Which pair of ratios does **not** demonstrate a proportional relationship?

   A. $\frac{8}{12}$ and $\frac{6}{9}$

   B. $\frac{9}{24}$ and $\frac{6}{16}$

   C. $\frac{12}{15}$ and $\frac{15}{20}$

   D. $\frac{27}{36}$ and $\frac{6}{8}$

6. Anil receives a 7% commission on computer sales. Last month his computer sales totaled $12,500. How much did Anil earn in commissions last month?

   _____

7. The lunch platter at one restaurant costs $7. Write an equation that shows the total cost, c, of p people buying the lunch platter.

_____

8. Maria paid $15.56 for a 4-pound roast beef. What is the unit price?

_____

9. Ian took out a 3-year loan for $1600. He has to pay 5% simple interest on the loan. How much interest will Ian have to pay?

_____

10. Which of the following tables has values that show a proportional relationship?

A.

a	1	2	3
b	1	4	9

B.

a	1	2	3
b	5	10	15

C.

a	1	2	3
b	3	5	7

D.

a	1	2	3
b	2	3	4

11. On a map, 3 cm represents 60 km. What is the constant of proportionality?

_____

12. The Millers' electric bill jumped from $84 in May to $105 in June. What is the percent of increase in their electric bill?

_____

13. Look at the figures below.

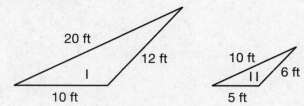

Do the corresponding sides form a proportional relationship?

_____

_____

14. A bucket is being filled at a rate of $\frac{5}{6}$ gallon in $\frac{1}{4}$ minute. What is the unit rate in gallons per minute?

_____

15. Marina and her mother each have a salad and a drink for lunch at a restaurant. A salad costs $4.75 and a drink costs $1.65. Marina's mother leaves a 15% tip. What is the amount of the tip?

_____

16. Graph $y = 3x$ on the coordinate plane. Identify the unit rate from the graph.

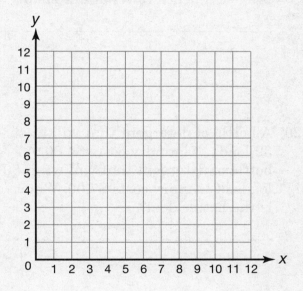

unit rate: _____

17. Madison estimated that there were 918 jelly beans in a jar. The actual number of jelly beans in the jar was 850. What was the percent error of Madison's estimate?

_____

18. What is the missing value in the following proportion?

    $\frac{8}{40} = \frac{12}{n}$

    A. 44

    B. 60

    C. 75

    D. 80

19. Jenna owns a souvenir shop at the beach. She marks up the price of all souvenirs by 40%. If beach towels cost her $8 each, what will be the price of a beach towel at her shop?

    _____

20. A bottle of detergent that will wash 32 loads of laundry costs $4.80. A bottle of detergent that will wash 50 loads of laundry costs $7.00. Which is the better deal?

    _____

21. Andrea bought an item at an online website. The item sold for $180 and Andrea was charged a $9 fee for shipping. What percent of the sale price was the shipping fee?

    _____

22. Look at the graph below.

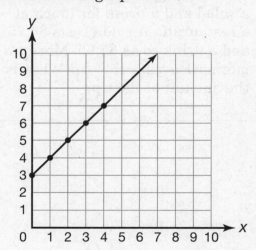

    Does the graph show a proportional relationship?

    _____

23. Lina took out a 4-year loan for $14,000 to buy a new car. She has to pay 4% simple interest on the loan. How much will Lina have paid in all after the 4 years?

_____

24. Look at the figures below.

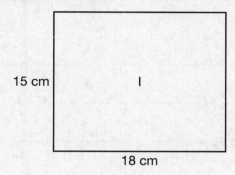

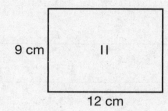

Do the corresponding sides of the figures form a proportional relationship?

_____

25. The price of a lawnmower went from $450 in May to $360 in September. What was the percent of decrease?

_____

27. Write an equation that represents the proportional relationship shown in the table.

x	6	4	2
y	3	2	1

_____

26. Alberto measured the length of a piece of wire to be 105 cm. The actual length of the wire was 120 cm. What was the percent error of Alberto's measurement?

_____

28. A washing machine normally sells for $650.

Part A
During a holiday sale, each washing machine is marked down 20%. How much will it cost to buy the washing machine during the sale?

_____

Explain how you found your answer.

_____

_____

_____

Part B
There is a delivery fee of $39 added to the sale price. What percent of the sale price is the delivery fee?

_____

Explain how you found your answer.

_____

_____

_____

Part C
The sales tax rate is 4%. How much tax will be paid on the washing machine during the sale? Do not include the delivery fee.

_____

Explain how you found your answer.

_____

_____

_____

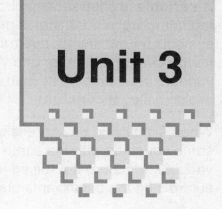

Unit 3

Expressions and Equations

You may be wondering how algebra is used in the real world. Chemists use equations to describe chemical reactions. Biologists use equations to analyze the growth of organisms. Engineers use inequalities to model real-world situations. Algebra is used to describe patterns and relationships of numbers.

In this unit, you will simplify and combine expressions. You will write linear equations and linear inequalities. Then you will use algebra to solve word problems involving linear equations and linear inequalities.

In This Unit

Simplifying and Combining Expressions

Writing Linear Equations

Using Algebra to Solve Word Problems (Linear Equations)

Writing Linear Inequalities

Using Algebra to Solve Word Problems (Linear Inequalities)

CCSS: 7.EE.1, 7.EE.2, 7.EE.3

Lesson 11: Simplifying and Combining Expressions

A **variable** is a letter or symbol used to represent a quantity. An **expression** is a mathematical phrase that is made up of variables and/or numbers, usually joined by symbols for addition or subtraction. A **term** of an expression is a number, a variable, or the product of a number and variable. When a term is the product of a number and a variable, the number is called the **coefficient** of the variable. A term without a variable is also called a **constant**.

When simplifying an expression, you can combine like terms. **Like terms** have the same variables with the same corresponding exponents. In **linear expressions** the variable terms are not raised to a power. You can combine like terms by adding or subtracting the coefficients only.

 Example

Simplify the following expression.
$8x + 5x$

Combine like terms.
$8x$ and $5x$ are like terms because they have the same variable, x.
$8x + 5x = 13x$

 Example

Simplify the following expression.
$9p + 8 - 4p - 5$

Use the commutative property of addition to change the order of the addends.
$9p + 8 - 4p - 5 = 9p - 4p + 8 - 5$

Use the associative property of addition to group the addends.
Combine like terms and combine the constant terms.
$(9p - 4p) + (8 - 5) = 5p + 3$

$9p + 8 - 4p - 5 = 5p + 3$

 TIP: You can also combine like terms by using the distributive property of multiplication over addition.

$8x + 5x = (8 + 5) \cdot x = 13x$

CCSS: 7.EE.1, 7.EE.2, 7.EE.3

To **expand** an expression means to multiply out the products in the expression and simplify. You can use the distributive property to expand an expression.

 Example

Expand the following expression.
$4(z - 7)$

Use the distributive property of multiplication over subtraction.
$4(z - 7) = 4 \cdot z - 4 \cdot 7$
$\qquad\qquad = 4z - 28$

$4(z - 7) = 4z - 28$

To **factor** an expression means to express it as a product in which one of the factors is an expression or rational number. You can use the distributive property to factor.

 Example

Factor the following expression.
$5y + 10$

Rewrite each term as a product. Find the greatest common factor (GCF) of 5 and 10. The GCF is 5.
$5y + 10 = 5 \cdot y + 5 \cdot 2$

Write as a product of factors.
$5 \cdot y + 5 \cdot 2 = 5(y + 2)$

$5y + 10 = 5(y + 2)$

 Example

Factor the following expression.
$12 - 8t$

Rewrite each term as a product. Find the greatest common factor of 12 and 8. The GCF is 4.
$12 - 8t = 4 \cdot 3 - 4 \cdot 2t$

Write as a product of factors.
$4 \cdot 3 - 4 \cdot 2t = 4(3 - 2t)$

$12 - 8t = 4(3 - 2t)$

 TIP: You can check your answer by using the distributive property to multiply the factors. In the example above, $4(3 - 2t) = 4 \cdot 3 - 4 \cdot 2t = 12 - 8t$.

Example

Write an equivalent expression for the following expression.
$8(y - 5) - 5y$

First, use the distributive property of multiplication over subtraction to expand the product in the expression.
$8(y - 5) - 5y = 8 \cdot y - 8 \cdot 5 - 5y$
$= 8y - 40 - 5y$

Next, use the commutative property of addition to rewrite the expression so that like terms are next to each other.
$8y - 40 - 5y = 8y - 5y - 40$

Then combine like terms by grouping the terms using the associative property of addition.
$8y - 5y - 40 = (8y - 5y) - 40$
$= 3y - 40$

$8(y - 5) - 5y$ and $3y - 40$ are equivalent expressions.

 Example

Write an equivalent expression for the following expression.
$-3(k + 6) + 5(k - 1)$

First, use the distributive property of multiplication to expand the products.
$-3(k + 6) + 5(k - 1) = -3 \cdot k + (-3) \cdot 6 + 5 \cdot k + 5 \cdot (-1)$
$= -3k - 18 + 5k - 5$

Next, use the commutative and associative properties of addition. Rewrite the expression so that like terms are next to each other.
$-3k - 18 + 5k - 5 = -3k + 5k - 18 - 5$
Then combine like terms.
$-3k + 5k - 18 - 5 = 2k - 23$

$-3(k + 6) + 5(k - 1)$ and $2k - 23$ are equivalent expressions.

CCSS: 7.EE.1, 7.EE.2, 7.EE.3

Different forms of an expression can show how quantities are related.

▷ **Example**

Write two expressions to show that decreasing by 15% is the same as multiplying by 0.85.

15% can be expressed as $\frac{15}{100}$ or 0.15.

If the original amount is represented by d, a decrease of 15% can be represented by $d - 0.15d$.

Simplify the expression.
$$d - 0.15d = (1 - 0.15)d$$
$$= 0.85d$$

The expressions $d - 0.15d$ and $0.85d$ show that decreasing by 15% is the same as multiplying by 0.85.

▷ **Example**

Carol is getting a 10% raise. Her current salary is $12.50 per hour. Write and evaluate an expression to find Carol's new salary.

10% can be expressed as $\frac{10}{100}$ or $\frac{1}{10}$.

The expression for Carol's new salary is $12.50 + \frac{1}{10}(12.50)$.

Simplify the expression.
$$12.50 + \frac{1}{10}(12.50) = 12.50 + 1.25 = 13.75$$

Carol's new salary will be $13.75 per hour.

 TIP: An increase of 10% can also be expressed as $x + 0.1x$, which is equal to $(1 + 0.1)x$, or $1.1x$.

CCSS: 7.EE.1, 7.EE.2, 7.EE.3

⬤ Practice

Directions: For questions 1 through 12, simplify each expression.

1. $7y + 2y$ _____

2. $-8d + 8d$ _____

3. $12a - 5a$ _____

4. $-5c - 6c$ _____

5. $4p - 11p$ _____

6. $2h - 9h + 5h$ _____

7. $11q + 8 - q$ _____

8. $-4k - 7 + 5k$ _____

9. $7m - 10 - 4m + 3$ _____

10. $-5n + 1 - 8n - 6$ _____

11. $8r - 9 - 2r - 1$ _____

12. $12 - 6z - 5 + 10z$ _____

13. $8x - 10x + x$

   A. $-2x$

   B. $-x$

   C. x

   D. $2x$

14. $6w - 12 - w + 11$

   A. $5w - 1$

   B. $5w + 1$

   C. $7w - 1$

   D. $7w + 1$

Directions: For questions 15 through 20, expand each expression.

15. $6(x + 2)$ _____

16. $4(3y - 1)$ _____

17. $-3(5p + 7)$ _____

18. $8(3 - x)$ _____

19. $7(1 - 6c)$ _____

20. $-5(4 - 3y)$ _____

Directions: For questions 21 through 24, factor each expression to its simplest form.

21. $3x + 12$ _____

22. $15 - 10k$ _____

23. $9a + 15$ _____

24. $-20 - 6b$ _____

25. $8n - 20$

   A. $2(4n - 10)$
   B. $4(2n - 5)$
   C. $8(n - 12)$
   D. $-12n$

26. $24 - 16r$

   A. $8r$
   B. $2(12r - 8)$
   C. $4(6 - 4r)$
   D. $8(3 - 2r)$

Directions: For questions 27 through 36, simplify each expression.

27. $4(a + 3) + 2a$ _____

28. $5(b - 2) - 7$ _____

29. $3(2h - 1) - h$ _____

30. $2(3p - 4) + 5$ _____

31. $10 - 2(m + 5)$ _____

32. $3(x - 1) + 4(x - 2)$ _____

33. $15 + 5(y - 4) + 5$ _____

34. $2(6 + n) + 5(n - 3)$ _____

35. $a + 2 + 3(2a - 1)$ _____

36. $6(p - 3) + 2(5p + 1)$ _____

37. $8 + 3(4x - 1)$

    A. $12x + 5$

    B. $12x + 7$

    C. $12x + 11$

    D. $12x - 11$

38. $-7(b + 1) + 8(b - 3)$

    A. $15b + 31$

    B. $15b - 31$

    C. $b + 31$

    D. $b - 31$

39. Write two expressions that show that increasing a quantity by 4% is the same as multiplying that quantity by 1.04.

_____ and _____

40. Write two expressions that show that decreasing a quantity by 10% is the same as multiplying that quantity by 0.9.

_____ and _____

41. Which of the following expressions has the same meaning as "increase n by 25%"?

A. $0.25n$

B. $0.75n$

C. $1.25n$

D. $1.75n$

42. What does the expression $0.65d$ represent?

A. increase d by 35%

B. decrease d by 35%

C. increase d by 65%

D. decrease d by 65%

43. Jonathan earns $850 per week. He is getting a 5% raise.

Write an expression that can be used to find Jonathan's new weekly salary.

_____

How much will Jonathan's new weekly salary be?

_____

44. Identify the property for each step used to write an equivalent expression for $2(x - 7) + 5x$. Then write the equivalent expression.

$2(x - 7) + 5x = 2x - 14 + 5x$ by the _____ property of _____ over _____.

$2x - 14 + 5x = 2x + 5x - 14$ by the _____ property of _____.

$2x + 5x - 14 = (2x + 5x) - 14$ by the _____ property of _____.

$(2x + 5x) - 14 = $ _____

Lesson 12: Writing Linear Equations

An **equation** is a mathematical sentence stating that two quantities are equal. You can write an equation to solve a word problem. The following list shows the operations that are described by words or phrases in word problems.

Addition: sum, more, more than, plus, increased by, gain, exceed
Subtraction: difference, less, less than, minus, decreased by, diminish
Multiplication: product, multiplied by, times, twice, triple, quadruple
Division: quotient, divided by, ratio, half, third, fourth, per

When the operation is addition or multiplication, the order in which you write the terms does not matter since these are commutative. When the main operation is subtraction or division, the order in which you write the terms does matter.

To write an equation for a word problem, first choose a variable to represent the unknown quantity. Then translate the words of the problem into an equation.

▶ **Examples**

Sentence	Equation
A number increased by twelve is twenty.	$n + 12 = 20$
Five times a number is 30.	$5n = 30$
Eleven less than a number is nine.	$n - 11 = 9$
One-fourth of a number is eight.	$\frac{1}{4}n = 8$
The product of a number and six is negative eighteen.	$6n = -18$
The difference between twice a number and three is nine.	$2n - 3 = 9$
A number divided by seven is equal to negative four.	$\frac{n}{7} = -4$
The sum of a number and four multiplied by $\frac{1}{2}$ is sixteen.	$\frac{1}{2}(n + 4) = 16$

CCSS: 7.EE.4

You may have to translate the information given in a word problem into an equation.

▷ Example

Samantha bought jewelry at a crafts fair. She bought a necklace for $8 and some bracelets for $5 each. She spent $23 in all. Write an equation that can be used to find the number of bracelets Samantha bought.

Identify the given information.
She bought a necklace for $8 and some bracelets for $5 each. She spent $23 in all.

You need to find how many bracelets Samantha bought.
Choose a variable to represent the unknown quantity.
Let b represent the number of bracelets Samantha bought.

$$8 + 5b = 23$$

● Practice

Directions: For questions 1 through 8, identify the part or parts of each sentence that indicates one or more operations. Then choose a variable to represent the unknown number and write the correct equation.

1. Four less than a number is equal to seventeen. _____

2. Ten times a number equals negative thirty. _____

3. A number increased by three is twenty-five. _____

4. Twice a number is twelve. _____

5. The quotient of a number and negative nine is six. _____

6. Five more than three times a number is thirty-eight. _____

7. The sum of a number and eleven, multiplied by one-third is thirteen. _____

8. The difference between triple a number and fifteen is twenty-one. _____

9. Kate swam 18 laps at practice. This is three-fourths the number of laps that Paco swam at practice. Write an equation that can be used to find the number of laps, *p*, that Paco swam.

_____

10. The Statue of Liberty is 151 ft tall, but with the pedestal and foundation it is 305 ft tall. Write an equation that can be used to find the height of the pedestal and foundation, *h*.

_____

11. James is 19 years old. He is three years older than twice his sister Meredith's age. Write an equation that can be used to find Meredith's age, *m*.

_____

12. Rosa ordered 3 pizzas. She paid $21.75 after using a $5-off coupon. Write an equation that can be used to find the regular price, *p*, of a pizza.

_____

13. Which mathematical equation represents the following sentence?

    Five less than one-sixth of a number is twelve.

    A. $5 - \frac{1}{6}x = 12$ C. $\frac{1}{6}x - 5 = 12$

    B. $12 - \frac{1}{6}x = 5$ D. $\frac{1}{6}x - 12 = 5$

14. After a carpenter cut a 9.5 inch-long piece of wood from a board, the remaining part of the board was 26.5 inches long. Which equation can be used to find the original length of the board?

    A. $b - 9.5 = 26.5$ C. $9.5b = 26.5$

    B. $26.5 - b = 9.5$ D. $b + 9.5 = 26.5$

15. Carmen jogged three more than half as many miles this week as last week. This week she jogged twelve miles. Write an equation that can be used to find how many miles Carmen jogged last week.

_____

    Explain how you knew what operations to use in the equation.

_____

CCSS: 7.EE.3, 7.EE.4.a

Lesson 13: Using Algebra to Solve Word Problems (Linear Equations)

To solve equations, use inverse operations to isolate the variable on one side of the equation. Addition and subtraction are inverse operations. Multiplication and division are also inverse operations. A two-step equation includes two operations. To solve two-step equations, you can use two inverse operations.

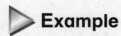 **Example**

Solve the following equation for y. $\qquad$ $4y + 3 = 27$

Since subtraction is the inverse of addition, subtract 3 from both sides of the equation.
$$4y + 3 = 27$$
$$4y + 3 - 3 = 27 - 3$$
$$4y = 24$$

Since division is the inverse of multiplication, divide both sides of the equation by 4.
$$\frac{4y}{4} = \frac{24}{4}$$
$$y = 6$$

To check your answer, substitute 6 for y in the original equation.
$$4y + 3 = 27$$
$$4 \cdot 6 + 3 = 27$$
$$24 + 3 = 27$$
$$27 = 27$$

Notice that you used multiplication and addition to check your answer. When you solved the equation, you used the inverse operations of multiplication and addition, which are division and subtraction.

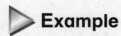 **Example**

Solve the following equation for n. $\qquad$ $2n - 3.8 = 6.4$

Since addition is the inverse of subtraction, add 3.8 to both sides of the equation.
$$2n - 3.8 = 6.4$$
$$2n - 3.8 + 3.8 = 6.4 + 3.8$$
$$2n = 10.2$$

Since division is the inverse of multiplication, divide both sides of the equation by 2.
$$2n = 10.2$$
$$\frac{2n}{2} = \frac{10.2}{2}$$
$$n = 5.1$$

Check your answer for reasonableness by estimating.

To the nearest whole number, 5.1 rounds to 5.

$2 \cdot 5 = 10$

3.8 rounds to 4. To the nearest whole number, 6.4 rounds to 6.

$10 - 4 = 6$

The answer is reasonable.

The solution is $n = 5.1$.

▷ **Example**

Solve the following equation for x.

$\frac{1}{2}(x + \frac{1}{4}) = \frac{7}{8}$

Use the distributive property of multiplication over addition to simplify the left side of the equation.

$$\frac{1}{2} \cdot x + \frac{1}{2} \cdot \frac{1}{4} = \frac{7}{8}$$
$$\frac{1}{2}x + \frac{1}{8} = \frac{7}{8}$$

Since subtraction is the inverse of addition, subtract $\frac{1}{8}$ from both sides of the equation.

$$\frac{1}{2}x + \frac{1}{8} = \frac{7}{8}$$
$$\frac{1}{2}x + \frac{1}{8} - \frac{1}{8} = \frac{7}{8} - \frac{1}{8}$$
$$\frac{1}{2}x = \frac{6}{8}$$
$$\frac{1}{2}x = \frac{3}{4}$$

Since division is the inverse of multiplication, divide both sides of the equation by $\frac{1}{2}$.

$$\frac{1}{2}x = \frac{3}{4}$$
$$\frac{1}{2}x \div \frac{1}{2} = \frac{3}{4} \div \frac{1}{2}$$
$$x = \frac{3}{4} \cdot \frac{2}{1}$$
$$x = \frac{6}{4} = 1\frac{1}{2}$$

The solution is $x = 1\frac{1}{2}$.

The following steps can be helpful when solving word problems.

▶ **Example**

Jeffrey wants to center a painting on a wall that is $11\frac{3}{4}$ feet wide. The painting is $2\frac{1}{4}$ feet wide. How far from each end of the wall should Jeffrey place the painting?

Step 1: **Take time to study the problem.** What is the problem asking?
How far from each end of the wall should Jeffrey place the painting?

Step 2: **Evaluate the information given.** Write down the necessary information.
Wall: $11\frac{3}{4}$ feet wide

Painting: $2\frac{1}{4}$ feet wide

Step 3: **Select a strategy for solving the problem.** Which operation or strategy will solve the problem?
Write an equation to represent this situation, where d is the distance from the end of the wall.

Step 4: **Set up the problem.**
Think: The distances from each end of the wall to each side of the painting plus the width of the painting should be equal to the width of the entire wall.
$d + 2\frac{1}{4} + d = 11\frac{3}{4}$

Step 5: **Do the math and check the answer.**

$d + d + 2\frac{1}{4} = 11\frac{3}{4}$ — Use the commutative property of addition to move the like terms together.

$2d + 2\frac{1}{4} = 11\frac{3}{4}$ — Combine like terms.

$2d + 2\frac{1}{4} - 2\frac{1}{4} = 11\frac{3}{4} - 2\frac{1}{4}$ — Subtract $2\frac{1}{4}$ from both sides of the equation.

$2d = 9\frac{1}{2}$ — Divide both sides of the equation by 2.

$d = 4\frac{3}{4}$

Check your answer by estimating.
The wall is about 12 feet wide and the painting is about 2 feet wide.

$12 - 2 = 10 \qquad \frac{1}{2} \cdot 10 = 5$

The painting should be placed about 5 feet from the end of each wall.
The estimate is close to the actual answer.

Jeffrey should place the painting $4\frac{3}{4}$ feet from the end of the wall.

▷ **Example**

Sharon received some money for her birthday from her grandparents. Then she received $45 for babysitting last week. Sharon deposited half of the money she had received in a savings account. If the deposit was for $55, how much did Sharon's grandparents give her for her birthday?

Step 1: **Take time to study the problem.** What is the problem asking?
How much did Sharon's grandparents give her for her birthday?

Step 2: **Evaluate the information given.** Write down the necessary information.
Amount of money received for babysitting: $45
Amount of money deposited: $55
Part of money deposited: $\frac{1}{2}$

Step 3: **Select a strategy for solving the problem.** Which operation or strategy will solve the problem?

Write an equation to represent this situation, where m is the money Sharon received from her grandparents.

Step 4: **Set up the problem.**

Think: Sharon deposited half of the money she had, and the amount deposited was $55.
$$\frac{1}{2}(m + 45) = 55$$

Step 5: **Do the math and check the answer.**

$$\frac{1}{2}(m + 45) = 55$$ Use the distributive property of multiplication over addition.

$$\frac{1}{2} \cdot m + \frac{1}{2} \cdot 45 = 55$$

$$\frac{1}{2}m + 22.50 = 55$$

$$\frac{1}{2}m + 22.50 - 22.50 = 55 - 22.50$$ Subtract 22.50 from both sides of the equation.

$$\frac{1}{2}m = 32.50$$ Divide both sides of the equation by $\frac{1}{2}$.

$$m = 65$$

Check your answer by substituting 65 into the original equation.
$$\frac{1}{2}(65 + 45) = 55$$
$$\frac{1}{2}(110) = 55$$
$$55 = 55$$

Sharon received $65 from her grandparents for her birthday.

Practice

Directions: For questions 1 through 10, solve the equation for the given variable.

1. $6 = 2x + 3$ _____

2. $-7 + \frac{1}{4}z = 0$ _____

3. $1.2 = 0.6m - 0.9$ _____

4. $-7a + 1 = 15$ _____

5. $3.6 = 1.8y - 5.4$ _____

6. $17b + 4 = -4\frac{1}{2}$ _____

7. $3s + 5 = -21$ _____

8. $-2p - 4 = 9$ _____

9. $18 - \frac{t}{20} = -9$ _____

10. $2(r + 9) = 24$ _____

11. $12 - 12n = -48$

    A. $n = -5$
    B. $n = -3$
    C. $n = 3$
    D. $n = 5$

12. $\frac{1}{2}(x - \frac{4}{5}) = \frac{3}{5}$

    A. $x = 1\frac{1}{10}$
    B. $x = 1\frac{1}{5}$
    C. $x = 2$
    D. $x = 2\frac{4}{5}$

Directions: For questions 13 through 18, write an equation to represent each word problem. Then, solve the equation to find the answer to the problem.

13. Michael is making a backdrop for the school play. He bought 4 rolls of material to use. He used 27 square feet of material for the backdrop, and had 5 square feet of material left over. How many square feet of material, s, are in each roll?

    equation: _____ $s =$ _____

14. Tom buys some shirts for $15 each. He has a coupon for $9 off the total price. If he pays $36, how many shirts, s, did he buy?

    equation: _____ $s =$ _____

15. Barb had $112 in her savings account. Then she deposited the same amount of money into her savings account in each of the next 4 months. Now she has $333 in her savings account. How much money, m, did Barb deposit each month?

    equation: _____ $m =$ _____

16. Nico bought a bag of jelly beans. He ate 15 jelly beans and then gave one-third of the remaining jelly beans to his brother. If he gave 25 jelly beans to his brother, how many jelly beans, j, were in the bag that Nico bought?

    equation: _____ $j =$ _____

17. Last week, Bob went canoeing on a river. The cost to rent a canoe was $20 plus $9.75 per hour. For how many hours, h, did Bob rent the canoe if he spent $78.50 in all?

    equation: _____ $h =$ _____

18. At the deli, sandwiches with 6 ounces of meat cost $3.89. You can order extra meat for $0.25 per ounce. If Greg's sandwich cost him $5.14, how much extra meat, m, did he order?

    equation: _____ $m =$ _____

19. The perimeter of a rectangle is equal to twice the sum of its length and its width.
    One rectangle has a length of $13\frac{1}{8}$ inches and a perimeter of $42\frac{1}{2}$ inches.
    Write an equation that can be used to find the width, *w*, of the rectangle.

    _____

    Explain how to find the answer. Justify each step in the equation.

    _____

    _____

    What is the width of the rectangle?

    _____

    Explain how you can check your answer by using estimation.

    _____

    _____

20. Lissette's age is 3 years less than twice the age of her sister Francesca. Lissette is 13.
    Write an equation that can be used to find Francesca's age, *f*.

    _____

    Explain how to find the answer. Justify each step in the equation.

    _____

    _____

    How old is Francesca?

    _____

    Explain how you can check your answer.

    _____

    _____

Lesson 14: Writing Linear Inequalities

An **inequality** is a mathematical sentence in which two expressions are joined by a symbol other than an equal sign. For example, an inequality may contain a greater than ($>$) or less than ($<$) symbol instead of an equal sign. Sometimes you will have to write an inequality to solve a problem. The following list shows how the symbols are described by words or phrases in word problems.

Greater than ($>$): greater than, more than
Greater than or equal to ($\geq$): at least, has a minimum value of, not less than
Less than ($<$): less than
Less than or equal to ($\leq$): at most, has a maximum of, no more than

The steps that you've used to write equations can be used to write inequalities, as well.

 Example

Sentence	Inequality
A number decreased by eight is less than fifteen.	$n - 8 < 15$
Four times a number is at least twenty.	$4n \geq 20$
The difference between half a number and six is at most ten.	$\frac{1}{2}n - 6 \leq 10$
The sum of five times a number and seven is more than eighteen.	$5n + 7 > 18$

You may have to translate the information given in a word problem into an inequality.

 Example

Landon has $51.75 in his wallet. He wants to buy some T-shirts that are on sale for $7.95 each and a pair of shorts for $8.95. Write an inequality that can be used to find the greatest number of T-shirts Landon can buy.

Identify the given information.
Landon has $51.75. He wants to buy some T-shirts that cost $7.95 each and a pair of shorts for $8.95.

You need to find the greatest number of T-shirts Landon can buy.
Choose a variable to represent the unknown quantity.

Let *t* represent the number of T-shirts Landon can buy. $7.95t$ represents the total cost of the shirts. Because the phrase used is "the greatest", the symbol described is $\leq$.

$7.95t + 8.95 \leq 51.75$

CCSS: 7.EE.4

 Practice

Directions: For questions 1 through 4, choose a variable to represent the unknown number and write the correct inequality.

1. A number multiplied by two, decreased by four, is greater than nine. _____

2. Seven-tenths of a number is at most forty-nine. _____

3. Five times a number, increased by six, is less than or equal to negative twenty.

   _____

4. The quotient of a number and four, decreased by $\frac{3}{8}$, is at least $\frac{-1}{8}$. _____

5. The cost to run an ad in a newspaper is $10 plus $0.25 per word. Write an inequality that can be used to find the maximum number of words Marianne can put in her ad if the most she can spend is $15.

   _____

6. A group of up to 40 people are going on a trip to Washington, DC. Some will travel in a van that holds 12 people, and the rest will buy train tickets. Write an inequality that can be used to find the number of train tickets that the group will need.

   _____

7. A used bookstore is having a sale. All paperback books are $0.75 and all hardcover books are $3. Ginny has $10. She wants to buy one hardcover book and as many paperback books as she can. Which inequality represents this situation, where p represents the number of paperback books Ginny can buy?

   A. $3 + 0.75p < 10$

   B. $3 + 0.75p \leq 10$

   C. $3 + 0.75p \geq 10$

   D. $3 + 0.75p > 10$

CCSS: 7.EE.3, 7.EE.4.b

Lesson 15: Using Algebra to Solve Word Problems (Linear Inequalities)

To solve inequalities, follow the same processes that apply to equations.

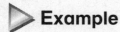 **Example**

Solve the inequality $\frac{x}{3} + 5 > 17$ for x.

$$\frac{x}{3} + 5 > 17$$

$$\frac{x}{3} + 5 - 5 > 17 - 5 \qquad \text{Subtract 5 from both sides.}$$

$$\frac{x}{3} > 12$$

$$3\left(\frac{x}{3}\right) > 3(12) \qquad \text{Multiply both sides by 3.}$$

$$x > 36$$

Check your solution by substituting any value that satisfies your solution for x in the original inequality. Then simplify. Since $39 > 36$, let $x = 39$.

$$\frac{39}{3} + 5 > 17$$

$$13 + 5 > 17$$

$$18 > 17$$

Since the substitution makes the inequality true, the solution is $x > 36$.

If you multiply or divide both sides of an inequality by a negative number, you have to switch the inequality sign to the opposite direction.

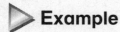 **Example**

Solve the inequality $-2x - 3 \leq 21$ for x.

$$-2x - 3 \leq 21$$

$$-2x - 3 + 3 \leq 21 + 3 \qquad \text{Add 3 to both sides.}$$

$$-2x \leq 24$$

$$\frac{-2x}{-2} \geq \frac{24}{-2} \qquad \text{Divide both sides by } -2 \text{ and switch the sign.}$$

$$x \geq -12$$

Check your solution. Since $0 > -12$, let $x = 0$.

$$-2 \bullet (0) - 3 \leq 21$$

$$-3 \leq 21$$

Since the substitution makes the inequality true, the solution is $x \geq -12$.

 TIP: If 0 is a value that satisfies your answer, it can be a useful number to substitute when checking your solution.

CCSS: 7.EE.3, 7.EE.4.b

You can graph the solution set of an inequality on a number line.

▷ **Example**

Graph the solution set for $5x - 4 < 36$.

$5x - 4 < 36$

$5x - 4 + 4 < 36 + 4$ Add 4 to both sides.

$5x < 40$

$\dfrac{5x}{5} < \dfrac{40}{5}$ Divide both sides by 5.

$x < 8$

Graph the solution set on a number line.

Put an open circle at 8 and draw a line pointing to the left.

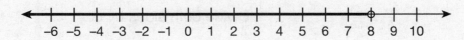

The number line shows the solution to the inequality.

▷ **Example**

Graph the solution set for $10 - \dfrac{x}{3} \leq 8$.

$10 - \dfrac{x}{3} \leq 8$

$10 - 10 - \dfrac{x}{3} \leq 8 - 10$ Subtract 10 from both sides.

$\dfrac{-x}{3} \leq -2$

$-3 \cdot \dfrac{-x}{3} \geq -3 \cdot -2$ Multiply both sides by -3 and switch the sign.

$x \geq 6$

Graph the solution set on a number line.

Put a closed circle at 6 and draw a line pointing to the right.

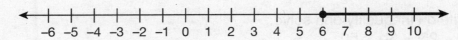

The number line shows the solution to the inequality.

◆ **TIP:** Use an open circle on the graph of an inequality that uses $<$ or $>$ to show that the number is not part of the solution set. Use a closed circle on the graph of an inequality that uses $\leq$ or $\geq$ to show that the number is part of the solution set.

Sometimes you need to interpret the results of an inequality to find the solution to a word problem.

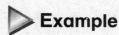 **Example**

Andrew has $40. He wants to buy a $15 CD and as many $6 posters as he can. What is the greatest number of posters Andrew can buy? Graph the solution set.

Write an inequality and solve.

$$6x + 15 \leq 40$$

$6x + 15 - 15 \leq 40 - 15$ Subtract 15 from both sides.

$$6x \leq 25$$

$\dfrac{6x}{6} \leq \dfrac{25}{6}$ Divide both sides by 6.

$$x \leq 4\tfrac{1}{6}$$

Graph the solution set on a number line.

Put a closed circle at $4\tfrac{1}{6}$ and draw a line pointing to the left.

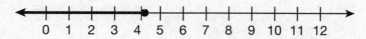

Since the number of posters must be a whole number, the greatest number of posters Andrew can buy is 4.

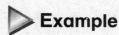 **Example**

Seena works part time selling newspapers over the phone. She is paid $25 per week plus $3 for each newspaper subscription she sells. She wants to earn at least $50 this week. What is the least number of newspaper subscriptions she needs to sell? Graph the solution set.

Write an inequality and solve.

$$25 + 3n \geq 50$$

$25 - 25 + 3n \geq 50 - 25$ Subtract 25 from both sides.

$$3n \geq 25$$

$\dfrac{3n}{3} \geq \dfrac{25}{3}$ Divide both sides by 3.

$$n \geq 8\tfrac{1}{3}$$

Graph the solution set on a number line.

Put a closed circle at $8\tfrac{1}{3}$ and draw a line pointing to the right.

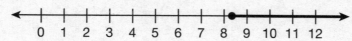

Since the number of subscriptions must be a whole number, the least number of subscriptions Seena must sell is 9.

116

Practice

Directions: For questions 1 through 10, solve the inequality for the given variable.

1. $4x + 15 \leq -3$ _____

2. $12 - 15x > 3$ _____

3. $2x - 4\frac{2}{3} \geq 3\frac{1}{2}$ _____

4. $-6a + 8 < 3$ _____

5. $\frac{5}{8}m - 7 \leq 8$ _____

6. $\frac{t}{4} - 7 \leq 1$ _____

7. $-2r - 4 > 44$ _____

8. $\frac{3}{5}s - 22 < 8$ _____

9. $2.5p + 3.9 \geq 8.1$ _____

10. $13 + 4n < 9$ _____

11. Which of the following values is in the solution set of the inequality $5x + 3 > 39$?

    A. 5

    B. 6

    C. 7

    D. 8

12. Which of the following values is in the solution set of the inequality $8 - 7n < -6$?

    A. -3

    B. -2

    C. 2

    D. 3

Directions: For questions 13 through 15, graph the solution set.

13. $3x - 11 < -17$

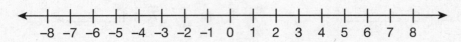

14. $\frac{1}{2}x + 4 \geq 1$

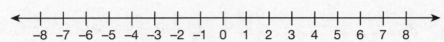

15. $-5x - 12 \leq 3$

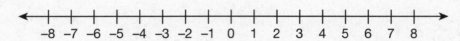

16. Frank bought two shirts that had the same price. He paid with $20 and received less than $3 in change. What is the minimum price that Frank could have paid for each shirt?

_____

17. Jonah wants to save at least $2,000 in a year. In the first three months of the year, he saved $650. What is the least average amount he must save for each of the remaining nine months of the year?

_____

Directions: For question 18, write and solve an inequality. Then graph the solution set on the number line.

18. Carl is having a party at his house. His mother is letting him spend at most $20 on snack food. He bought drinks for $4 and wants to buy some frozen pizzas that cost $5 each. What is the greatest number of pizzas Carl can buy?

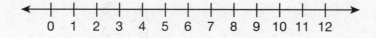

_____

19. Cindy plans to join a gym. A gym membership costs $350 for the year and each exercise class costs an additional $18. Cindy does not want to spend more than $1000 on the gym for the year.

    Write an inequality that can be used to find the maximum number of exercise classes Cindy can take in one year.

    _____

    Solve the inequality.

    _____

    How many exercise classes can Cindy take? Explain your answer.

    _____

    _____

    _____

20. Jeb wants to buy a DVD recorder that costs $127. He is going to save money for the DVD recorder by walking his neighbor's dog for $14 per week. Jeb has already saved $22.

    Write an inequality that can be used to find the minimum number of weeks Jeb must walk the dog in order to buy the DVD recorder.

    _____

    Solve the inequality.

    _____

    How many weeks must Jeb walk the dog in order to have enough money to buy the DVD recorder?

    _____

    Graph the solution set on the number line.

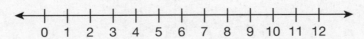

119

Unit 3 Practice Test

Directions: For questions 1 through 5, simplify each expression.

1. $12y - 4y$ _____

2. $3h - 10h + 6h$ _____

3. $4(3p - 1) + 9$ _____

4. $13 - 5(m + 4)$ _____

5. $5(2b - 1) - b$ _____

Directions: For questions 6 and 7, factor each expression into simplest form.

6. $3x - 27$ _____

7. $16 - 12k$ _____

8. Which of the following expressions has the same meaning as "increase by 20%"?

   A. $0.2n$

   B. $0.8n$

   C. $1.2n$

   D. $1.8n$

9. Which mathematical inequality represents the following sentence?

   Five more than one-third of a number is less than or equal to negative four.

   A. $\frac{1}{3}x + 5 > -4$

   B. $\frac{1}{3}x + 5 \geq -4$

   C. $\frac{1}{3}x + 5 < -4$

   D. $\frac{1}{3}x + 5 \leq -4$

10. In June, Jasmine earned $1200 in commission. In July, her commission decreased by 7%.

Write an expression that can be used to find Jasmine's commission in July.

_____

How much was Jasmine's commission in July?

_____

Directions: For questions 11 through 18, use the variable *n* to represent the unknown number and write the correct equation or inequality.

11. Six less than twice a number is equal to ninety.

_____

12. The quotient of a number and eight, increased by $\frac{3}{4}$, is at least $\frac{-1}{4}$.

_____

13. The sum of a number and seven, multiplied by four, is thirty-six.

_____

14. Three-eighths of a number is at most seventy-two.

_____

15. One-half of a number, decreased by one-fifth, is nine-tenths.

_____

16. Five more than three times a number is less than negative ten.

_____

17. A number divided by three, increased by eight, is fifty-nine.

_____

18. The difference between a number multiplied by seven and fourteen is greater than nine.

_____

19. Michael bought 4 pairs of socks. He paid $7.16 after using a $2-off coupon. Write an equation that can be used to find the regular price, *p*, of a pair of socks.

_____

20. Identify the property for each step used to simplify the expression $3(x - 4) + 7x$. Then write the equivalent simplified expression.

    $3(x - 4) + 7x = 3x - 12 + 7x$ by the _____ property of _____ over _____.

    $3x - 12 + 7x = 3x + 7x - 12$ by the _____ property of _____.

    $3x + 7x - 12 = (3x + 7x) - 12$ by the _____ property of _____.

    $(3x + 7x) - 12 =$ _____

Directions: For questions 21 through 30, solve the equation or inequality for the given variable.

21. $19 = 3x + 7$ _____

22. $-8 + \frac{1}{5}z > 0$ _____

23. $1.7 = 0.4m - 0.7$ _____

24. $-6a + 2 \leq 20$ _____

25. $5.2 = 1.4y - 1.8$ _____

26. $6(2x + 3) < -9$ _____

27. $11 - 12b = 3$ _____

28. $\frac{1}{4}x - 3\frac{1}{2} \geq 6\frac{1}{4}$ _____

29. $-2(c + 6) = 1$ _____

30. $\frac{5}{6}m - 5 \leq 10$ _____

31. Which of the following values is **not** in the solution set of the inequality $7x - 5 > 29$?

    A. 4
    B. 5
    C. 6
    D. 7

32. Solve the following equation for x.
    $\frac{1}{4}(x - \frac{3}{8}) = \frac{1}{8}$

    A. $x = \frac{1}{2}$

    B. $x = \frac{3}{4}$

    C. $x = \frac{7}{8}$

    D. $x = 1\frac{1}{8}$

Directions: For questions 33 through 35, graph the solution set.

33. $6x + 7 > -17$

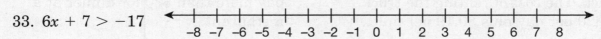

34. $\frac{1}{2}x - 3 \le -1$

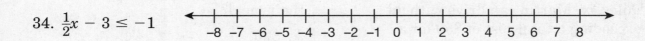

35. $-2x - 6 \ge 4$

```
←——+——+——+——+——+——+——+——+——+——+——+——+——+——+——+——+——+→
   -8 -7 -6 -5 -4 -3 -2 -1  0  1  2  3  4  5  6  7  8
```

36. Amelia paid $225 to have her television fixed. The repair shop charged $35 per hour for repairs and $85 for parts.

Write an equation that can be used to find how many hours, h, it took to repair Amelia's television.   _____

Solve the equation.

_____

37. Matthew had $4.10 in his pocket. He bought 3 notebooks. Now he has $0.35 in his pocket.

Write an equation that can be used to find how much each notebook, n, cost.   _____

Solve the equation.

_____

38. Dwayne brought his trading card collection to school in the morning. At lunch, his friend Jason gave him 8 more trading cards. After school, Dwayne bought some more trading cards which doubled the number of cards he had. At the end of the day, Dwayne had 62 trading cards.

Write an equation to find how many trading cards, t, Dwayne had in the morning.   _____

Solve the equation.

_____

39. Martin is planning to rent a small truck. The cost of renting the truck is $20 per day plus $0.19 per mile.

    Write an inequality that can be used to find the maximum number of miles, m, Martin can drive without exceeding a total cost of $50 for a one-day rental.

    _____

    What is the maximum number of miles?

    _____

40. A county fair charges $5 for admission and $0.75 for each ride. Zena is going to the fair with $15.

    Write an inequality that can be used to find the greatest number of rides, r, that Zena can go on.

    _____

    What is the greatest number of rides Zena can go on?

    _____

41. Sean and Emily do not want to pay more than $30 for dinner at a restaurant.

    They plan to leave a tip of 20%. What is the most they can spend on their meal?

    A. $20

    B. $24

    C. $25

    D. $36

42. A tennis player received a $100 gift certificate to a sporting goods store. She used it to buy a pair of sneakers for $75 and as many cans of tennis balls as possible. The cans of tennis balls cost $2.75 each. What is the greatest number of cans of tennis balls that she could buy?

    _____

    Graph the solution set.

    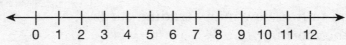

43. Nicole bought a bag of balloons. She blew up 25 balloons for a party and then gave one-third of the remaining balloons to her sister. She gave 10 balloons to her sister.

Part A
Write an equation that can be used to find how many balloons, b, were in the bag that Nicole bought.

_____

Part B
Solve the equation.

_____

Part C
Explain how you can check your answer.

_____

_____

_____

44. Ed is ordering magazines online. Each magazine costs $6. Shipping will cost an additional $4 for the entire order.

Part A
Write an inequality that can be used to find the number of magazines, n, Ed can order if he spends no more than $42.

_____

Part B
What is the greatest number of magazines Ed can order? Explain your answer.

_____

_____

Part C
Graph the solution set.

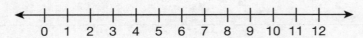

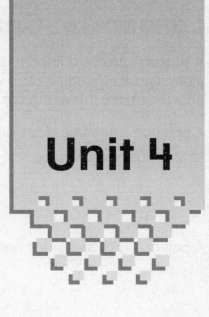

Unit 4

Geometry

Geometry appears in the real world in countless ways. For example, engineers and architects use many different shapes in the designs of bridges and buildings. They use properties of angles to add strength as well as visual appeal. They also make drawings and models before building large structures.

In this unit, you will construct geometric figures. You will identify the relationships between angles and find the measures of unknown angles in figures. You will find the areas of polygons. You will solve problems involving scale drawings of geometric figures. You will also find the circumference and area of circles. You will describe the two-dimensional figures that result from slicing three-dimensional figures. Finally, you will find the volume and surface area of three-dimensional objects.

In This Unit

Constructing Geometric Figures

Angle Relationships

Interior and Exterior Angles

Areas of Objects

Scale Drawings

Circles

Cross-Sections of Three-Dimensional Figures

Surface Areas and Volumes of Objects

CCSS: 7.G.2

Lesson 16: Constructing Geometric Figures

A **plane figure** is another name for a two-dimensional figure. **Polygons** are closed plane figures formed by line segments that are connected at points called **vertices**. All polygons have three or more line segments, or sides.

A polygon with three sides and three angles is called a **triangle**.

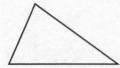

A polygon with four sides and four angles is called a **quadrilateral**.

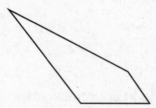

Other common polygons include the **pentagon**, **hexagon**, and **octagon**.

pentagon   hexagon   octagon

You can draw a figure based on the properties you are given.

▶ **Example**

Draw a quadrilateral that has only one pair of parallel sides.

Step 1: Draw a pair of parallel segments of different lengths.

Step 2: Draw segments that connect the endpoints of the parallel segments.

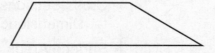

If you are given specific lengths of sides or measures of angles for a figure, you can use a ruler and a protractor to construct the figure.

 Example

Construct a triangle that has two sides measuring 3 in. and 4 in., and an angle between these sides measuring 20°.

Step 1: Use a ruler to draw a line segment that is 4 in. long.

Step 2: Use a protractor to construct a 20° angle using an endpoint of the line segment as the vertex.

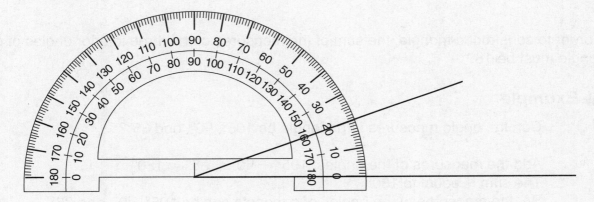

Step 3: Use a ruler to draw a line segment from the vertex that is 3 in. long.

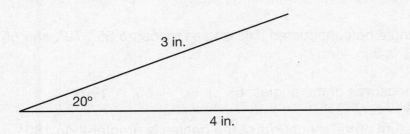

Step 4: Use a ruler to complete the triangle.

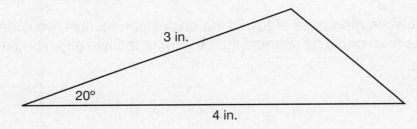

129

Not all sets of lengths and/or measures can form a polygon. For example, in order to construct a triangle, the sum of the lengths of the two shorter sides must be greater than the length of the longest side.

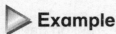 **Example**

Can the lengths of the sides of a triangle be 5 cm, 8 cm, and 14 cm?

Add the lengths of the two shorter sides: $5 + 8 = 13$
Compare the sum to the length of the longest side: $13 < 14$
The sum of the lengths of the two shorter sides is less than the length of the longest side.
So, the lengths of the sides of a triangle cannot be 5 cm, 8 cm, and 14 cm.

In order to construct a triangle, the sum of the measures of the three interior angles of a triangle must be 180°.

 Example

Can the angle measures of a triangle be 105°, 40°, and 35°?

Add the measures of the angles: $105° + 40° + 35° = 180°$
The sum is equal to 180°.
So, the measures of the angles of a triangle can be 105°, 40°, and 35°.

 Example

Can a triangle be constructed if its angles measure 55°, 70°, and 65°, and one of its sides is 6 in. long?

Add the measures of the angles: $55° + 70° + 65° = 190°$
$190° > 180°$
Since the sum of the measures of the angles is greater than 180°, a triangle cannot be constructed with these angle measures.

 TIP: If you know the measures of two of the angles of a triangle, you can find the measure of the third angle by subtracting the sum of the two given angle measures from 180°.

CCSS: 7.G.2

Some sets of lengths and/or measures can form several different figures.

 Example

Construct a triangle whose angles measure 45°, 60°, and 75°.

Step 1: Use a ruler to draw a line segment of any length, since no specific length is given.

_____

Step 2: Use a protractor to construct a 45° angle using an endpoint of the line segment as a vertex.

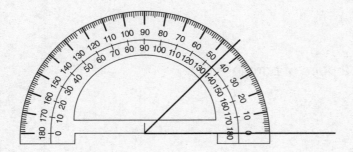

Step 3: Use a protractor to construct a 60° angle using the other endpoint of the line segment as another vertex.

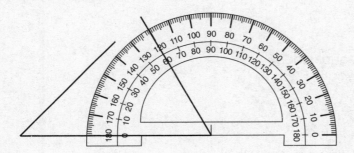

Step 4: Extend the line segments you drew in Steps 2 and 3 until they meet. You can use your protractor to verify the measure of the third angle.

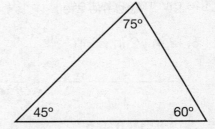

 TIP: Two figures that have the same shape but not the same size, are called **similar figures**. The measures of the corresponding angles of two similar figures are the same.

131

⬤ Practice

Directions: For questions 1 through 5, draw the geometric shape.

1. a triangle that has a 90° angle

2. a triangle that has 3 equal sides

3. a quadrilateral that has 2 pairs of parallel sides

4. a quadrilateral that has 4 equal sides but no right angles

5. a pentagon that has 5 equal sides

Directions: For questions 6 through 11, tell whether a triangle can be constructed using the given measures.

6. 2 in., 2 in., 3 in.

   _____

7. 3 ft, 15 ft, 9 ft

   _____

8. 4 yd, 6 yd, 5 yd

   _____

9. 85°, 20°, 65°

   _____

10. 52°, 73°, 55°

    _____

11. 10 cm, 115°, 75°

    _____

CCSS: 7.G.2

Directions: For questions 12 and 13, construct a triangle using the given measures. Then state whether there can be more than one triangle constructed with these measures.

12. a triangle whose angles measure 30°, 55°, and 95°

_____

13. a triangle that has two angles measuring 70° and 45° and the side between the angles measuring 2 in.

_____

14. There can be more than one triangle that has two sides measuring 8 cm and 5 cm and an angle with a measure of 50°.

    Sketch and label the possible triangles.

    How many triangles can you construct with sides measuring 8 cm and 5 cm and an angle with a measure of 50°? Explain your answer.

    _____

    _____

    _____

Lesson 17: Angle Relationships

An **angle** is formed by two rays that share a common **endpoint**. The common endpoint (*B*) is called the **vertex**. The rays ($\overrightarrow{BA}$ and $\overrightarrow{BC}$) are called **sides** of the angle.

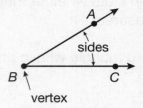

The angle shown can be named ∠*B*, ∠*ABC*, ∠*CBA*, or ∠1. If three letters are used to name an angle, the middle letter names the vertex.

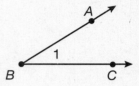

Angles Formed by Intersecting Lines

Intersecting lines may form special angle relationships.

Two angles that share a common vertex and a common ray between them but do not have any interior points in common are called **adjacent angles**.

∠*WXY* and ∠*YXZ* are adjacent angles.

Complementary angles are two angles that have a sum of 90°. The measure of an angle is denoted by the letter *m*.

$m\angle WXY + m\angle YXZ = 90°$

If complementary angles are adjacent to each other, then the two rays that the angles do not share form a right angle and are **perpendicular** to each other. This is indicated by a small square in the corner of the angle.

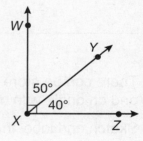

Supplementary angles are two angles that have a sum of 180°.

$m\angle ABD + m\angle DBC = 180°$

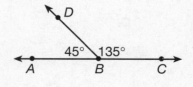

Vertical angles are the angles formed by two intersecting lines. Vertical angles are congruent, which means that they have the same measure.
The symbol for congruent is ≅.
∠1 ≅ ∠3   ∠2 ≅ ∠4

CCSS: 7.G.5

 Example

What is $m\angle JKL$ in the diagram below?

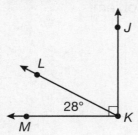

You can set up the following equation because $\angle JKL$ and $\angle LKM$ are complementary angles.

$x + 28 = 90$
$x = 62$
Therefore, $m\angle JKL = 62°$.

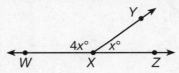 **Example**

What is $m\angle WXY$ and $m\angle YXZ$ in the diagram below?

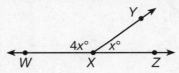

You can set up the following equation because $\angle WXY$ and $\angle YXZ$ are supplementary angles.

$4x + x = 180$
$5x = 180$
$x = 36$
So, $4x = 4 \cdot 36 = 144$.
Therefore, $m\angle WXY = 144°$ and $m\angle YXZ = 36°$.

 Example

What are the values of x and y in the diagram below?

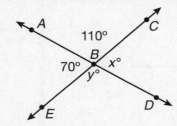

Because $\angle ABC$ and $\angle EBD$ are vertical angles, their measures are equal.
Because $\angle ABE$ and $\angle CBD$ are vertical angles, their measures are equal.
Therefore, $x = 70$ and $y = 110$.

◯ **Practice**

Directions: For questions 1 through 5, use the information given to write an equation that can be used to solve the problem. Then solve the problem.

1. What is $m\angle SQR$?

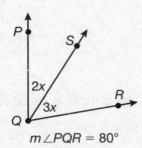

$m\angle PQR = 80°$

equation: _____ $m\angle SQR =$ _____

2. What is $m\angle ABC$?

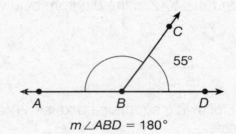

$m\angle ABD = 180°$

equation: _____ $m\angle ABC =$ _____

3. If the measure of $\angle DBE$ is 40° and angles EBC and ABD are congruent, what is $m\angle EBC$?

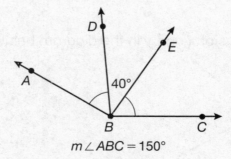

$m\angle ABC = 150°$

equation: _____ $m\angle EBC =$ _____

136

CCSS: 7.G.5

4. What is $m\angle ABC$?

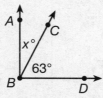

$\angle ABC$ and $\angle CBD$ are complementary.

equation: _____ $m\angle ABC =$ _____

5. What is $m\angle ABC$?

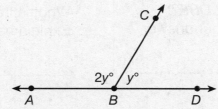

$\angle ABC$ and $\angle CBD$ are supplementary.

equation: _____ $m\angle ABC =$ _____

Directions: Use the figure below to answer questions 6 and 7.

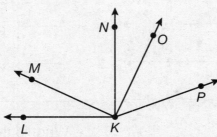

$m\angle LKN$ and $m\angle MKO$ are 90°.

6. What is the measure of $\angle LKO$ if the measure of $\angle NKO$ is 25°?

    A.   25°

    B.   55°

    C.   90°

    D. 115°

7. What is the measure of $\angle PKO$ if the measure of $\angle PKM$ is 135°?

    A.   25°

    B.   35°

    C.   45°

    D. 65°

CCSS: 7.G.5

8. Use the figure below to answer the questions.

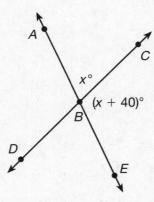

What is the measure of ∠DBE? Explain how you found your answer.

_____

_____

_____

What is the measure of ∠ABD? Explain how you found your answer.

_____

_____

_____

9. Use the figure below to answer the questions.

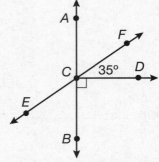

$\overleftrightarrow{AB}$ intersects $\overleftrightarrow{EF}$ at point C.

What is the measure of ∠ACF? Explain how you found your answer.

_____

_____

_____

What is the measure of ∠BCE? Explain how you found your answer.

_____

_____

_____

What is the measure of ∠ACE? Explain how you found your answer.

_____

_____

_____

CCSS: 7.G.5

Lesson 18: Interior and Exterior Angles

When two parallel lines are cut by a **transversal**, certain angle relationships are formed.

The symbol ∥ is used to indicate that lines are parallel. In the following diagram, *l* ∥ *m*.

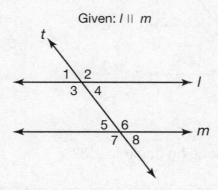

Given: *l* ∥ *m*

Interior angles are the angles that are between the two parallel lines cut by a transversal.
∠3, ∠4, ∠5, and ∠6 are interior angles.

Exterior angles are the angles that are outside the two parallel lines cut by a transversal.
∠1, ∠2, ∠7, and ∠8 are exterior angles.

Alternate interior angles are interior angles that are on opposite sides of a transversal. When formed by parallel lines cut by a transversal, alternate interior angles are congruent.
∠3 ≅ ∠6   ∠4 ≅ ∠5

Alternate exterior angles are exterior angles that are on opposite sides of a transversal. When formed by parallel lines cut by a transversal, alternate exterior angles are congruent.
∠1 ≅ ∠8   ∠2 ≅ ∠7

Corresponding angles are angles that are on the same side of a transversal and in corresponding positions with respect to the parallel lines. Each pair of corresponding angles includes one interior angle and one exterior angle. When formed by parallel lines cut by a transversal, corresponding angles are congruent.
∠1 ≅ ∠5   ∠2 ≅ ∠6   ∠3 ≅ ∠7   ∠4 ≅ ∠8

TIP: When parallel lines are cut by a transversal, all of the acute angles formed are congruent and all of the obtuse angles formed are congruent.

▷ **Example**

In the figure below, lines *l* and *m* are parallel.

Given: *l* ∥ *m*

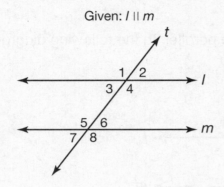

What is the corresponding angle to ∠3?

Corresponding angles lie on the same side of the transversal, in corresponding positions.

∠3 lies on the left side of the transversal below line *l*.

The corresponding angle lies on the left side of the transversal below line *m*.

∠7 lies on the left side of the transversal below line *m*.

So, ∠7 is the corresponding angle to ∠3.

 Example

If the measure of ∠7 is 50° in the figure above, what is the measure of ∠4?

∠7 and ∠3 are corresponding angles.

Corresponding angles have the same measure.

So, $m\angle 3 = 50°$.

∠3 and ∠4 are supplementary angles.

So, $m\angle 3 + m\angle 4 = 180°$.

$50 + m\angle 4 = 180$

The measure of ∠4 is 130°.

CCSS: 7.G.5

▷ Example

In the figure below, lines *m* and *n* are parallel.

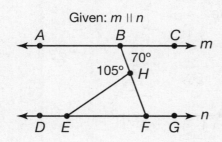

What is the measure of ∠*DEH*?

∠*CBF* and ∠*BFE* are alternate interior angles. Therefore, they are congruent. So, *m*∠*BFE* = 70°.

∠*BHE* and ∠*FHE* are supplementary angles and add up to 180°. ∠*BHE* is given to be 105°.

105 + *m*∠*FHE* = 180

m∠*FHE* = 75°

The sum of the measures of the angles of a triangle is 180°.

m∠*HFE* + *m*∠*FHE* + *m*∠*FEH* = 180

70 + 75 + *m*∠*FEH* = 180

145 + *m*∠*FEH* = 180

m∠*FEH* = 35°

∠*FEH* and ∠*DEH* are supplementary angles and add up to 180°. ∠*FEH* is equal to 35°.

35 + *m*∠*DEH* = 180

So, the measure of ∠*DEH* is 145°.

⬤ Practice

Directions: Use the following figure to answer questions 1 through 10. Do not use a

protractor, as the angles are not drawn to scale.

Given: $r \parallel s$

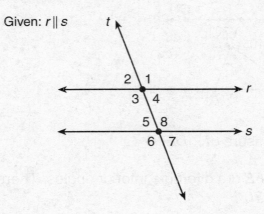

1. List all pairs of corresponding angles. _____

2. List all pairs of alternate exterior angles. _____

3. List all pairs of vertical angles. _____

4. List all pairs of alternate interior angles. _____

5. List two pairs of supplementary angles. _____

6. If $m\angle 3 = 100°$, what is $m\angle 1$? _____

7. If $m\angle 4 = 65°$, what is $m\angle 8$? _____

8. If $m\angle 2 = 75°$, what is $m\angle 1$? _____

9. If $m\angle 6 = 105°$, what is $m\angle 2$? _____

10. If $m\angle 5 = 70°$, what is $m\angle 4$? _____

CCSS: 7.G.5

Directions: Use the figure below to answer questions 11 and 12.

Given: *m* ∥ *n*

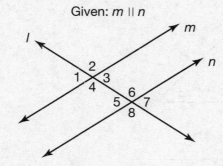

11. If $m\angle 1$ is represented by $5x$ and $m\angle 2$ is represented by $7x$, what is the value of x?

    A.   12

    B.   15

    C.   75

    D. 105

12. If $m\angle 3$ is 75°, what is $m\angle 8$?

    A.   75°

    B.   85°

    C.   95°

    D. 105°

Directions: For questions 13 through 15, use the figure below to write an equation that can be used to find the angle. Then find the measure of the angle.

Given: *p* ∥ *q*

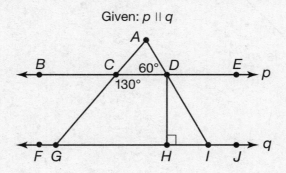

13. $m\angle ACD$ equation: _____ $m\angle ACD$ = _____

14. $m\angle CAD$ equation: _____ $m\angle CAD$ = _____

15. $m\angle HDI$ equation: _____ $m\angle HDI$ = _____

Directions: For questions 16 through 18, find the value of *x*.

16.

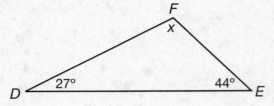

x = _____

17.

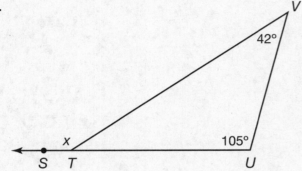

x = _____

18.   Given: $\overrightarrow{AF} \parallel \overline{CB}$

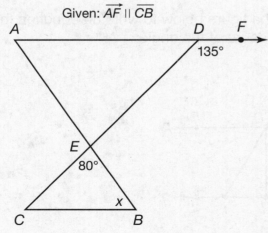

x = _____

CCSS: 7.G.5

19. Use the figure below to answer the questions.

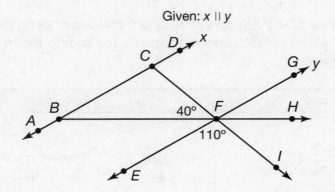

What is the measure of ∠GFH? Explain how you found your answer.

_____

_____

What is the measure of ∠ABF? Explain how you found your answer.

_____

_____

20. Use the figure below to answer the question.

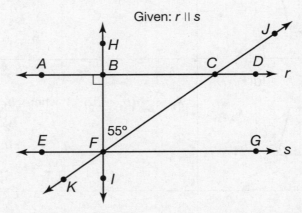

What is the measure of ∠EFK? Explain how you found your answer.

_____

_____

Lesson 19: Areas of Objects

Area (A) is the measure of the region inside a two-dimensional figure. Area is measured in square units. The following table shows formulas for finding the areas of several types of two-dimensional figures.

Figure	Formula	
triangle	$A = \frac{1}{2}bh$	where b = base length
		h = height
square	$A = s^2$	where s = length of each side
rectangle	$A = lw$	where l = length
		w = width
parallelogram	$A = bh$	where b = base length
		h = height
trapezoid	$A = \frac{1}{2}h(b_1 + b_2)$	where b_1 = base 1 length
		b_2 = base 2 length
		h = height
regular polygon	$A = \frac{1}{2}ap$	where a = length of apothem
		p = perimeter

CCSS: 7.G.6

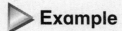 **Example**

What is the area of the following trapezoid?

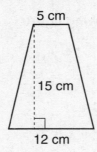

5 cm

15 cm

12 cm

Substitute the values into the formula and solve.

$A = \frac{1}{2} h(b_1 + b_2)$

$= \frac{1}{2} \cdot 15 \cdot (5 + 12)$

$= 127.5$

The area of the trapezoid is 127.5 cm^2.

A regular polygon is a polygon in which all the sides are equal in measure and all the angles are congruent. You can find the area of a regular polygon if you know the perimeter and the length of the **apothem**, the distance from the midpoint of one side of the polygon to its center.

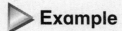 **Example**

What is the area of the following regular hexagon?

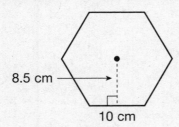

8.5 cm

10 cm

Find the perimeter of the polygon.
There are 6 sides and each side is 10 cm long.
$p = 6 \cdot 10 \text{ cm} = 60 \text{ cm}$

Substitute the values into the formula and solve.

$A = \frac{1}{2}ap$

$= \frac{1}{2} \cdot 8.5 \cdot 60$

$= 255$

The area of the regular hexagon is 255 cm^2.

You can find the area of an irregular polygon by breaking it into triangles.

▷ Example

What is the area of the following figure?

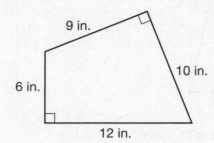

Draw a diagonal to divide the figure into two right triangles.

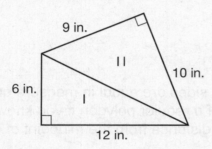

Step 1: Find the area of triangle I.

$$A = \frac{1}{2}bh$$
$$= \frac{1}{2} \cdot 12 \cdot 6$$
$$= 36$$

Step 2: Find the area of triangle II.

$$A = \frac{1}{2}bh$$
$$= \frac{1}{2} \cdot 9 \cdot 10$$
$$= 45$$

Step 3: Add the areas of the triangles.

$$A = 36 + 45$$
$$= 81$$

The area of the irregular quadrilateral is 81 in.²

CCSS: 7.G.6

The area of a composite figure can be found by first breaking it into common figures. Next, find and sum the areas of these common figures.

▷ Example

What is the area of the following figure?

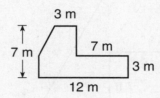

Step 1: Separate the figure into common figures.
The figure can be separated into a trapezoid and a rectangle.

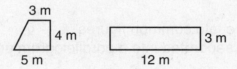

Step 2: Find the area of each figure.

Area of trapezoid

$$A = \frac{1}{2} h(b_1 + b_2)$$

$$= \frac{1}{2} \cdot 4 \cdot (3 + 5)$$

$$= 16$$

Area of rectangle

$$A = lw$$

$$= 12 \cdot 3$$

$$= 36$$

Step 3: Add the areas.

$$A = 16 + 36$$

$$= 52$$

The area of the figure is 52 m^2.

▷ **Example**

Ted is building a deck that wraps around a corner of his house. The shape of the deck is shown below.

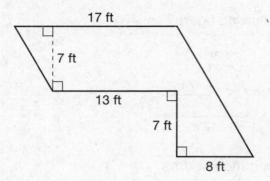

What is the area of the deck?

Step 1: Separate the figure into common figures.
The figure can be separated into a parallelogram and a trapezoid.

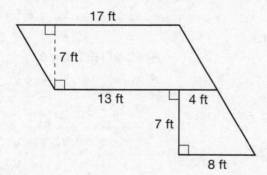

Step 2: Identify the measurements for each figure.
The base of the parallelogram is 17 ft and its height is 7 ft.
The height of the trapezoid is 7 ft. The bases of the trapezoid are 8 ft and 4 ft (17 ft − 13 ft).

Step 3: Find the area of each figure.

Area of parallelogram

$A = bh$

$= 17 \cdot 7$

$= 119$

Area of trapezoid

$A = \frac{1}{2} h(b_1 + b_2)$

$= \frac{1}{2} \cdot 7 \cdot (8 + 4)$

$= 42$

Step 4: Add the areas of the figures.
$A = 119 + 42$
$= 161$

The area of the deck is 161 ft^2.

CCSS: 7.G.6

⬤ **Practice**

Directions: For questions 1 and 2, find the area of each plane figure. Circle the figure with the greater area.

1.

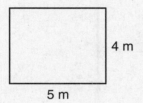

4 m
5 m

5 m
10 m

A = _____

A = _____

2.

18 yd
6 yd

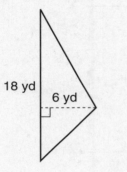

5 yd
9 yd

A = _____

A = _____

Directions: For questions 3 through 9, find the area of each figure.

3.

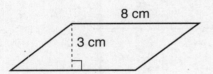

8 cm
3 cm

A = _____

4.

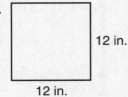

12 in.
12 in.

A = _____

151

5.

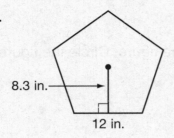

8.3 in. →

12 in.

A = _____

6.

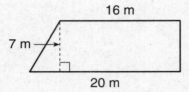

16 m

7 m →

20 m

A = _____

7.

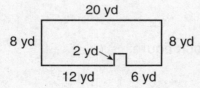

36 ft

12 ft

45 ft

A = _____

8.

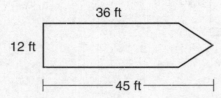

20 yd

8 yd   8 yd

2 yd →

12 yd   6 yd

A = _____

9.

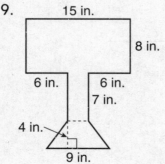

15 in.

8 in.

6 in.   6 in.

7 in.

4 in. →

9 in.

A = _____

CCSS: 7.G.6

10. A rectangular playground has an area of 7125 square yards. The width of the playground is 75 yards. What is the length of the playground?

    A. 65 yards

    B. 75 yards

    C. 85 yards

    D. 95 yards

11. What is the area of the figure?

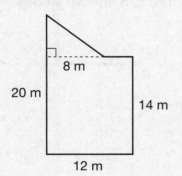

    A. 168 m²

    B. 192 m²

    C. 216 m²

    D. 240 m²

12. A square has a perimeter of 60 inches. What is the area of the square?

    A. 225 square inches

    B. 240 square inches

    C. 900 square inches

    D. 3600 square inches

13. What is the area of the figure?

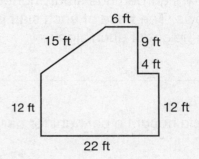

    A. 318 ft²

    B. 372 ft²

    C. 390 ft²

    D. 490 ft²

14. Mrs. Allen is buying a carpet for a rectangular room that measures 10 feet wide and 21 feet long. How many square feet of carpet will Mrs. Allen need?

_____

15. Yoshi bought new tiles for a square-shaped kitchen that measures 11 feet on each side. The tiles cost $1.50 per square foot. How much will Yoshi pay for the tiles?

_____

16. Dory's garden has small triangular signs that state the name of each vegetable she grows. The base of each sign is 4 inches and the area is 6 square inches. What is the height of each sign?

_____

17. Alicia bought a pennant for her school's football team.

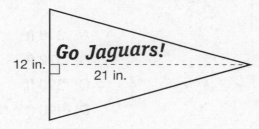

12 in.
Go Jaguars!
21 in.

What is the area of the pennant?

_____

18. A stop sign is in the shape of a regular octagon with a side length of 5.5 inches and an apothem of 6.5 inches.

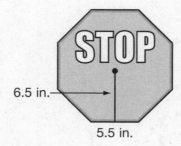

6.5 in.

5.5 in.

What is the area of the stop sign?

_____

CCSS: 7.G.6

19. What is the area of the figure? Show your work.

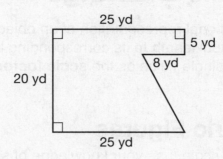

25 yd

5 yd

8 yd

20 yd

25 yd

Explain how you found your answer.

_____

_____

20. Leah and Matthew are building kites for a kite competition. Their kites are shown below.

Leah's kite   **Matthew's kite**

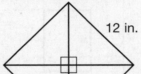

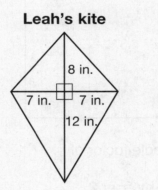

8 in.

7 in.   7 in.

12 in.

12 in.

12 in.

What is the area of Leah's kite? Explain how you found your answer.

_____

_____

What is the area of Matthew's kite? Explain how you found your answer.

_____

_____

Whose kite has the larger area? How much larger is it?

_____

155

Lesson 20: Scale Drawings

A **scale drawing** is a proportional representation of an object. The **scale** of a drawing is the constant ratio of each actual length to its corresponding length in the drawing. This scale can be expressed in a single value as the **scale factor.**

Drawing Geometric Figures

To draw geometric figures to scale, use your knowledge of similar figures and proportional relationships. The ratios of corresponding side lengths of similar figures are equal to the scale factor, so the scale factor indicates how much larger or smaller to make each side length in the scale drawing. Scaling of geometric figures preserves angle measures, so the corresponding angles are congruent.

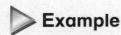 **Example**

Draw a rectangle that is 3 units wide and 4 units high.

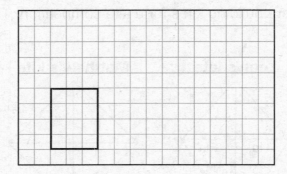

Draw another rectangle using a scale factor of $\frac{2}{1}$.

Multiply both the width and height by $\frac{2}{1}$, or 2.

new width: $3 \cdot 2 = 6$ new height: $4 \cdot 2 = 8$

Find a starting point and draw the new width and height. Remember that all of the sides of the new figure will meet at the same angles as the corresponding sides of the original figure.

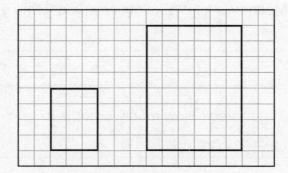

156

Example

Compare the area of the new rectangle to the area of the original rectangle in the previous example. How does the ratio of their areas relate to the scale factor?

The new rectangle is 6 units wide and 8 units high, so its area is 6 • 8, or 48 square units.

The original rectangle is 3 units wide and 4 units high, so its area is 3 • 4, or 12 square units.

Write a ratio that compares the areas in simplest form.
$$\frac{48}{12} = \frac{4}{1}$$

Since $\frac{4}{1} = (\frac{2}{1})^2$, the ratio of the areas is the square of the scale factor.

If two figures are similar, the ratio of their areas is the square of the scale factor.

 Example

The triangles shown below are similar figures. What is the ratio of the area of the scale triangle to the area of the original triangle?

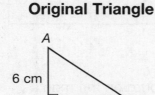

Original Triangle   **Scale Triangle**

Find the scale factor.
A pair of corresponding sides are $\overline{AB}$ and $\overline{DE}$. The ratio of $\overline{DE}$ to $\overline{AB}$ is $\frac{8}{6}$ or $\frac{4}{3}$.

Find the area of each triangle.

$\triangle ABC$

$A = \frac{1}{2}bh$

$\quad = \frac{1}{2} \cdot 9 \cdot 6$

$\quad = 27$

$\triangle DEF$

$A = \frac{1}{2}bh$

$\quad = \frac{1}{2} \cdot 12 \cdot 8$

$\quad = 48$

Write a ratio in simplest form that compares the areas.
$$\frac{48}{27} = \frac{16}{9}$$

The ratio of the areas is $\frac{16}{9}$, which is the square of the scale factor, $\frac{4}{3}$.

Using Scales on Maps

Maps are common examples of scale drawings. The distances on a map are proportional to the actual distances.

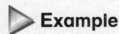 **Example**

On the following map, what is the approximate distance from Martin to Greensburg?
(The scale shows that 1 inch = 30 miles.)

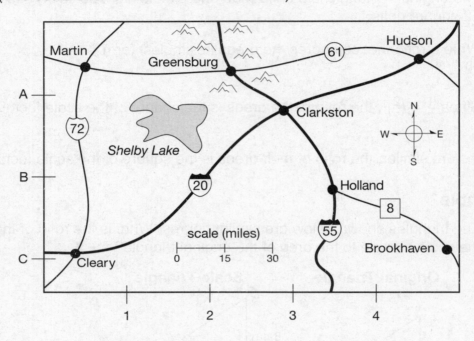

Measure the distance from Martin to Greensburg on the map.

$1\frac{1}{2}$ inches = 1.5 inches

Set up and solve a proportion to find the actual distance.

$$\frac{1.5 \text{ inches}}{x \text{ miles}} = \frac{1 \text{ inch}}{30 \text{ miles}}$$

$$1.5 \cdot 30 = 1 \cdot x$$
$$x = 45$$

The distance from Martin to Greensburg is about 45 miles.

CCSS: 7.G.1

⬤ **Practice**

Directions: For questions 1 through 3, make a scale drawing of the figure using the given scale factor. Then write the ratio of the areas.

1. scale factor: $\frac{1}{2}$

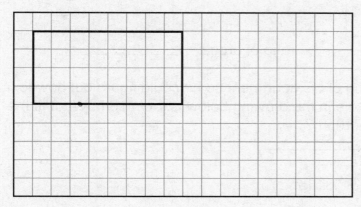

ratio of areas: _____

2. scale factor: $\frac{2}{3}$

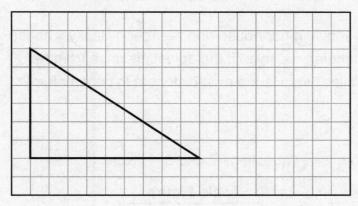

ratio of areas: _____

3. scale factor: $\frac{5}{3}$

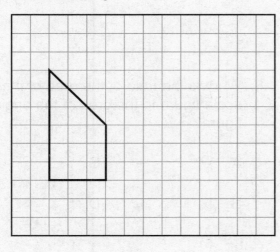

ratio of areas: _____

4. Re-draw the rectangle below with a scale factor of $\frac{2}{3}$. Give the ratio of the areas.

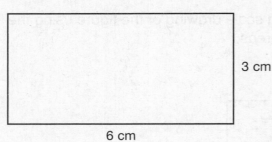

3 cm

6 cm

ratio of areas: _____

5. Alejandro made a scale drawing of his apartment so he could figure out how to rearrange the furniture. The scale in the drawing below is 1 cm : 3 ft. What is the actual length of Alejandro's bedroom?

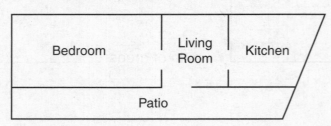

_____

6. Renata drew an accurate map showing her house and her friend Becky's house. The scale on the map is 1 centimeter $= \frac{1}{2}$ mile. If the actual distance from her house to Becky's house is $2\frac{1}{2}$ miles, what is the map distance, in centimeters?

_____

Directions: Use the figure below to answer questions 7 and 8.

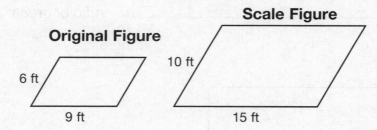

Scale Figure

Original Figure

10 ft

6 ft

9 ft

15 ft

7. What is the scale factor for the parallelograms?

    A. $\frac{9}{25}$ C. $\frac{3}{5}$

    B. $\frac{4}{9}$ D. $\frac{5}{3}$

8. What is the ratio of the area of the scale figure to the area of the original figure?

    A. $\frac{25}{9}$ C. $\frac{3}{5}$

    B. $\frac{4}{9}$ D. $\frac{2}{3}$

Directions: Use the floor plan below to answer questions 9 through 14. Use a ruler to measure the dimensions.

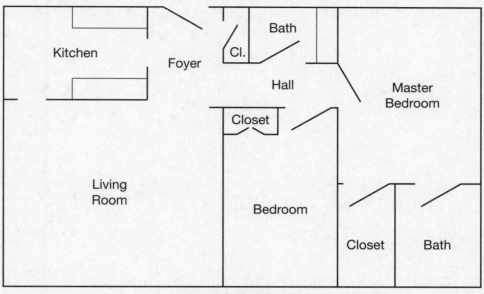

Scale: 1 inch = 8 feet

9. What is the actual length of the kitchen? _____

10. What is the actual width of the kitchen? _____

11. What is the actual area of the kitchen in square feet? _____

12. What is the actual length of the master bedroom, including the bath and closet?

_____

13. What is the actual width of the master bedroom? _____

14. What is the actual area of the master bedroom in square feet, including the bath and the closet?

_____

15. The actual dimensions of a rectangular porch are 16 feet long by 12 feet wide.

    Use a scale of 1 inch = 4 feet to make a scale drawing of the porch.

    Use a scale of 1 cm = 2 feet to make a scale drawing of the porch.

CCSS: 7.G.4

Lesson 21: Circles

A **circle** is a closed figure in a plane that consists of all points that are an equal distance from the **center** point. The name of a circle is given by its center. Here is circle *O*.

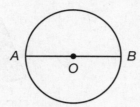

A **radius (*r*)** is any line segment from the center of the circle to a point on the circle. The radius is half as long as the diameter. $\overline{OA}$ and $\overline{OB}$ are radii of circle *O*.

A **diameter (*d*)** is any line segment that passes through the center and has both endpoints on the circle. The diameter is twice as long as the radius. $\overline{AB}$ is a diameter of circle *O*.

The **circumference (*C*)** is the distance around the circle.

For all circles, the ratio of the circumference to the diameter of the circle ($\frac{C}{d}$) is the same number. The number cannot be expressed exactly as a decimal or fraction. The number is called **pi** and is represented by the symbol **π**. The value of pi is approximately 3.14 or $\frac{22}{7}$.

$C \div d = \pi$ or $\frac{C}{d} = \pi$

Multiply both sides by *d*.

$\frac{C}{d} \cdot d = \pi \cdot d$

$C = \pi d$

▷ Example

Find the circumference of a circle having a diameter of 9 inches.

$C = \pi d$
$\approx 3.14 \cdot 9$
≈ 28.26

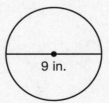

The circumference of the circle is about 28.26 inches.

▷ **TIP:** The length of the radius is half the length of the diameter, or $d = 2r$. Substituting 2*r* for *d* in the formula $C = \pi d$ gives the equivalent formula $C = 2\pi r$.

The formulas for the circumference of a circle and the area of a parallelogram can be used to show how the formula for the area of a circle was developed.

Divide a circle into sectors.

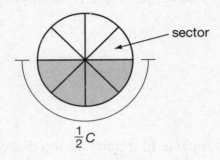

sector

$\frac{1}{2}C$

Since $C = 2\pi r$, $\frac{1}{2}C = \pi r$.

Separate the sectors and rearrange them side by side into a parallelogram.

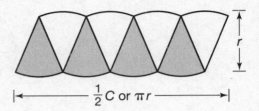

r

$\frac{1}{2}C$ or πr

The base of the parallelogram is about $\frac{1}{2}$ the circumference. The height of the parallelogram is the radius of the circle.

$$A = bh \approx \pi r \cdot r \approx \pi r^2$$

The formula for the area of a circle is $A = \pi r^2$.

▶ **Example**

Find the area of a circle having a radius of 4 inches.

$A = \pi r^2$

$\approx 3.14 \cdot 4^2$

$\approx 3.14 \cdot 16$

≈ 50.24 square inches

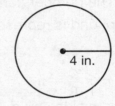

4 in.

The area of the circle is about 50.24 square inches.

CCSS: 7.G.4

 Practice

Directions: Use the following circle to answer questions 1 and 2.

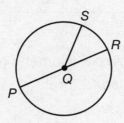

1. Name a line segment that is a radius of circle *Q*. _____

2. Name a line segment that is a diameter of circle *Q*. _____

3. If the diameter of a circle is 10 cm, what is its radius?

   A. 2.5 cm

   B. 5 cm

   C. 20 cm

   D. 25 cm

4. If the radius of a circle is 18 ft, what is its diameter?

   A. 4.5 ft

   B. 9 ft

   C. 13.5 ft

   D. 36 ft

5. A wheel has a diameter of 28 inches. What is the approximate distance around the outside of the wheel? Use $\frac{22}{7}$ for π.

   A. 44 in.

   B. 88 in.

   C. 176 in.

   D. 616 in.

6. A round clock has a radius of 5 cm. What is the approximate area of the clock? Use 3.14 for π.

   A. 31.4 cm^2

   B. 78.5 cm^2

   C. 314 cm^2

   D. 785 cm^2

7. A circular tablecloth is 5 feet in diameter. What is the approximate distance around its border? Use 3.14 for π.

_____

8. A pizzeria has free delivery within a 7-kilometer radius of the restaurant. What is the size of the free delivery area? Use $\frac{22}{7}$ for π.

_____

9. The diameter of a dinner plate is 10 inches. What is the approximate area of the plate? Use 3.14 for π.

_____

10. The radius of a circle is 7 inches.

    What is the circumference of the circle? Use $\frac{22}{7}$ for π. _____

    Suppose the radius is doubled to 14 inches. What is the circumference of this circle?

    _____

    Explain how doubling the radius affects the circumference.

    _____

11. The radius of a circle is 6 inches.

    What is the area of the circle? Use 3.14 for π. _____

    Suppose the radius is doubled to 12 inches. What is the area of this circle?

    _____

    Explain how doubling the radius affects the area.

    _____

Lesson 22: Cross-Sections of Three-Dimensional Figures

A **solid** is another name for a three-dimensional shape. The dimensions of a solid are its length, its width, and its height.

The plane figures that make up the boundaries of a solid are called its **faces**. Faces intersect to form **edges**. The point of intersection of three or more edges is called a **vertex**.

The **net** of a solid is a two-dimensional representation that shows how the figure appears when it is unfolded on a flat surface.

A prism has two parallel and congruent **bases**. The other faces of a prism are rectangles.

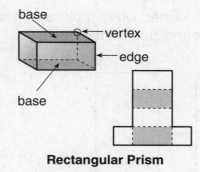

Rectangular Prism

A pyramid has only one base, and its faces are triangles that meet at a common vertex.

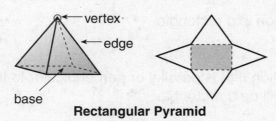

Rectangular Pyramid

CCSS: 7.G.3

A plane that intersects a solid forms a two-dimensional figure called a **cross-section**. The figure below shows a plane intersecting a **right rectangular prism**. The plane is parallel to the bases.

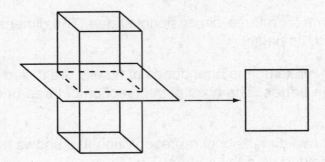

The cross-section is a square.

 Example

What cross-section is formed when a plane intersects a right rectangular prism so that it is perpendicular to the bases?

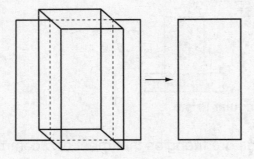

The cross-section is a rectangle.

TIP: Any cross-section that is parallel or perpendicular to the faces of a right rectangular prism will be a rectangle.

CCSS: 7.G.3

 Example

What cross-section is formed when a plane intersects a right rectangular pyramid so that it is parallel to the base?

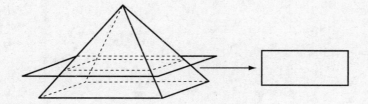

The cross-section is a rectangle.

 Example

What cross-sections are formed when a plane intersects a right rectangular pyramid so that it is perpendicular to the base?

There is more than one possible cross-section.

The plane is perpendicular to the base through the vertex.

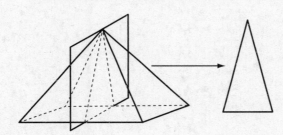

The cross-section is a triangle.

The plane is perpendicular to the base, but not through the vertex.

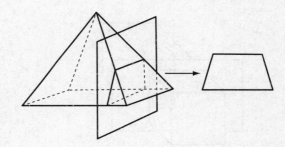

The cross-section is a trapezoid.

● Practice

Directions: For questions 1 through 4, find the shape of the cross-section that will be formed by the intersecting plane.

1.

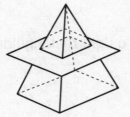

_____

2.

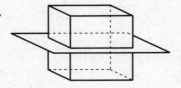

_____

3.

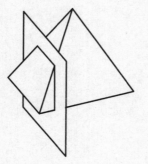

_____

4.

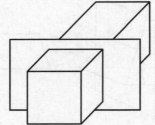

_____

5. Draw the cross-section formed by an intersecting plane that is parallel to the base.

6. Use the figure below to answer the questions.

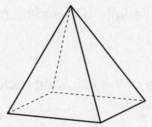

Draw the cross-section formed by an intersecting plane that is parallel to the base.

Draw the cross-section formed by a plane that is perpendicular to the base and intersects through the vertex.

Draw the cross-section formed by a plane that is perpendicular to the base and does not intersect through the vertex.

Lesson 23: Surface Areas and Volumes of Objects

Surface Area

Surface area (*S.A.*) is the total area of the outside surfaces of a three-dimensional figure. Surface area is measured in square units. You can use the following formula to find the surface area of a rectangular prism:

S.A. = 2(*lw* + *lh* + *wh*) where *l* = length, *w* = width, and *h* = height

 Example

What is the surface area of the following rectangular prism?

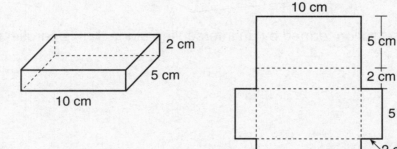

Substitute the values into the formula and solve.

$$S.A. = 2(lw + lh + wh)$$
$$= 2(10 \cdot 5 + 10 \cdot 2 + 5 \cdot 2)$$
$$= 2(50 + 20 + 10)$$
$$= 2(80)$$
$$= 160$$

The surface area of the rectangular prism is 160 cm².

If you need help visualizing how the formula works, refer to the net of the figure. Sum the areas of all 6 rectangles that make up the sides of the prism to find the surface area.

There are two rectangles that are 10 cm by 5 cm, two rectangles that are 10 cm by 2 cm, and two rectangles that are 5 cm by 2 cm.

$$S.A. = (10 \cdot 5) + (10 \cdot 5) + (10 \cdot 2) + (10 \cdot 2) + (5 \cdot 2) + (5 \cdot 2)$$
$$= 50 + 50 + 20 + 20 + 10 + 10$$
$$= 160$$

A cube is a right rectangular prism whose faces all have the same area. You can use the following formula to find the surface area of a cube:

$S.A. = 6e^2$ where e = length of an edge of the cube

 Example

Leona is going to wrap a box that is shaped like a cube. The box has a side length of 4 inches. What is the least amount of wrapping paper that Leona needs?

Substitute the values into the formula and solve.
$$S.A. = 6e^2$$
$$= 6 \cdot 4^2$$
$$= 6 \cdot 16$$
$$= 96$$

The surface area of the gift box is 96 in.^{2}.
Leona needs at least 96 in.^{2} of wrapping paper.

To find the surface area of a composite figure, add the surface areas of the individual figures, but then subtract out the areas of the surfaces where the figures meet.

 Example

What is the surface area of the following figure?

Find the surface area of the cube.
Each edge is 6 in.
$$S.A. = 6e^2$$
$$= 6 \cdot 6^2$$
$$= 6 \cdot 36$$
$$= 216$$
The surface area of the cube is 216 in.^{2}.

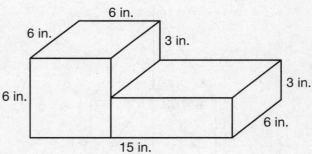

Find the surface area of the rectangular prism.
The length is 15 − 6, or 9 in., the width is 6 in., and the height is 3 in.

$S.A. = 2(lw + lh + wh)$

$= 2(9 \cdot 6 + 9 \cdot 3 + 6 \cdot 3)$

$= 2(54 + 27 + 18)$

$= 2(99)$

$= 198$

The surface area of the rectangular prism is 198 in.^{2}.

Add the surface areas of the figures, but subtract out the areas of the surfaces where the cube and rectangular prism intersect. The cube and rectangular prism meet at a rectangle which is 6 in. long and 3 in. wide. Its area is 18 in.^{2}. It must be subtracted twice because it appears on both figures.

$S.A. = 216 \text{ in.}^2 + 198 \text{ in.}^2 - 2(18 \text{ in.}^2) = 378 \text{ in.}^2$

Volume

Volume (*V*) is the amount of space that a solid takes up. Volume is measured in cubic units. You can use the following formula to find the volume of a prism:
$V = Bh$ where B = area of the base and h = height

▷ Example

What is the volume of the following prism?

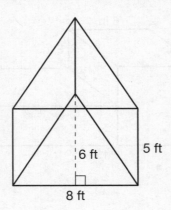

6 ft 5 ft

8 ft

The base is a triangle, so use the formula for the area of a triangle for *B*.
Substitute the values into the formula.

$V = Bh$

$= \frac{1}{2} \cdot 8 \cdot 6 \cdot 5$

$= 120$

The volume of the prism is 120 ft^3.

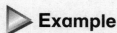

In a rectangular prism, the base is a rectangle, so the formula becomes $V = lwh$ where l = length, w = width, and h = height.

Since all the edges of a cube have the same length, the formula for the volume of a cube is $V = e^3$ where e = length of an edge.

To find the volume of a composite figure, add the volumes of the individual figures that make up the composite figure.

▶ **Example**

What is the volume of the following figure?

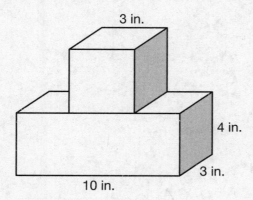

3 in.

4 in.

3 in.

10 in.

Find the volume of the cube.
$V = e^3$
$\quad = 3^3$
$\quad = 3 \cdot 3 \cdot 3$
$\quad = 27$
The volume of the cube is 27 in.^{3}.

Find the volume of the rectangular prism.
$V = lwh$
$\quad = 10 \cdot 3 \cdot 4$
$\quad = 120$

The volume of the rectangular prism is 120 in.^{3}.

Add the volumes.
$V = 27$ in.^{3} + 120 in.^{3} = 147 in.^{3}

Practice

Directions: For questions 1 through 4, find the surface area of the figure.

1.

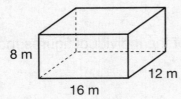

8 m

16 m

12 m

S.A. = _____

2.

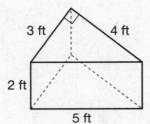

3 ft   4 ft

2 ft

5 ft

S.A. = _____

3.

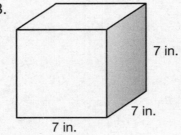

7 in.

7 in.

7 in.

S.A. = _____

4.

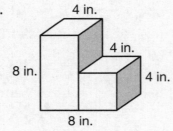

4 in.

4 in.

8 in.

4 in.

8 in.

S.A. = _____

5. The length, width, and height of a cereal box are 20 cm, 7 cm, and 28 cm. What is the least amount of cardboard that can be used to make the box?

_____

6. How many square feet of wood are needed to make a box that is 4 feet long, 2 feet high, and $2\frac{1}{2}$ feet wide?

_____

Directions: Use the figures below to answer questions 7 and 8.

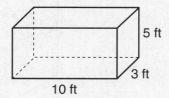

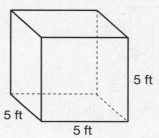

7. What is the surface area of the rectangular prism?

    A.   95 ft^2

    B. 150 ft^2

    C. 180 ft^2

    D. 190 ft^2

8. How much greater is the surface area of the rectangular prism than the surface area of the cube?

    A. They have the same surface area.

    B. 25 ft^2

    C. 30 ft^2

    D. 40 ft^2

CCSS: 7.G.6

Directions: For questions 9 through 11, find the volume of the figure.

9.

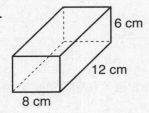

6 cm

12 cm

8 cm

$V = $ _____

10.

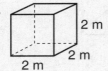

2 m

2 m

2 m

$V = $ _____

11.

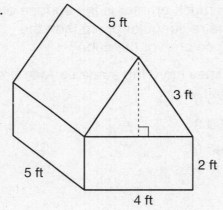

5 ft

3 ft

5 ft

2 ft

4 ft

$V = $ _____

12. A paperweight is in the shape of a prism whose base is a regular hexagon.

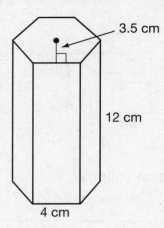

3.5 cm

12 cm

4 cm

What is the volume of the paperweight? _____

13. Tika is sculpting a clay figure in the shape of a rectangular prism.

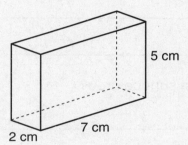

5 cm

7 cm

2 cm

What is the volume of clay that Tika will need to make the prism? _____

14. Sy is going to mail a glass ornament to his sister. He wants to use the box that will hold the greater volume of packing material.

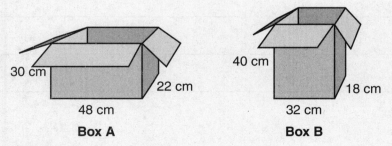

30 cm

22 cm

48 cm

Box A

40 cm

18 cm

32 cm

Box B

Which box has the greater volume? _____

15. An aquarium is 12 in. wide, 30 in. long, and 15 in. high.
What is the volume of the aquarium? _____

16. Use the figures to answer the questions below.

Which two boxes have the same volume?

_____

Explain how you found your answer.

_____

_____

_____

Which box has the greatest surface area?

_____

_____

Explain how you found your answer.

_____

_____

_____

Unit 4 Practice Test

1. If the radius of a circle is 15 ft, what is its diameter? _____

2. Draw a quadrilateral that has exactly one pair of parallel sides.

3. Construct a triangle that has two angles measuring 80° and 65°, and the side between the angles measuring 3 cm.

4. Which set of measurements can represent the lengths of the sides of a triangle?

   A. 4 cm, 8 cm, 6 cm

   B. 7 cm, 5 cm, 13 cm

   C. 9 cm, 3 cm, 3 cm

   D. 8 cm, 5 cm, 3 cm

5. If the diameter of a circle is 12 cm, what is its radius?

   A. 3 cm

   B. 6 cm

   C. 24 cm

   D. 36 cm

6. If the measure of $\angle 1$ is x and the measure of $\angle 2$ is $x + 50$, what is the measure of $\angle 4$?

   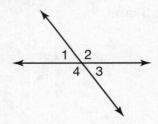

   A. 50°

   B. 55°

   C. 110°

   D. 115°

7. A square has a perimeter of 48 inches. What is the area of the square?

   A. 144 square inches

   B. 192 square inches

   C. 576 square inches

   D. 2304 square inches

8. Which set of measures could **not** represent the measures of the angles of a triangle?

   A. 75°, 35°, 70°

   B. 120°, 35°, 25°

   C. 85°, 55°, 50°

   D. 90°, 15°, 75°

9. A round table has a diameter of 35 inches. What is the approximate circumference of the table? Use $\frac{22}{7}$ for π.

   A. 55 in.

   B. 77 in.

   C. 110 in.

   D. 3850 in.

10. A circular plate has a radius of 11 cm. What is the area of the plate rounded to the nearest whole number? Use 3.14 for π.

_____

For questions 11 through 13, use the figure below to write an equation that can be used to find the angle measure. Then find the measure of the angle.

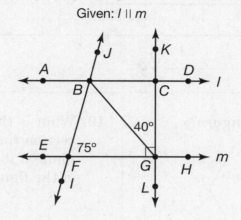

Given: *l* ∥ *m*

11. *m∠EFB*   equation: _____   *m∠EFB* = _____

12. *m∠EGB*   equation: _____   *m∠EGB* = _____

13. *m∠FBG*   equation: _____   *m∠FBG* = _____

14. What is the area of the figure?

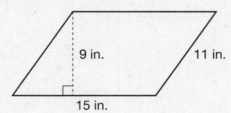

15. Draw a triangle that has 2 sides that are congruent.

16. Re-draw the figure below with a scale factor of $\frac{2}{3}$. Then write the ratio of the new area to the old area.

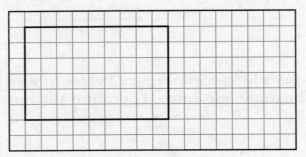

ratio of areas: _____

17. What is the area of the figure?

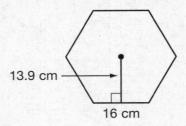

13.9 cm

16 cm

_____

18. Alana keeps her sweaters in a storage box under her bed. The box is 30 in. long, 18 in. wide, and 7 in. high. What is the volume of the storage box?

_____

19. What is the shape of the cross-section that will be formed by the plane if it passes through the vertex of the figure?

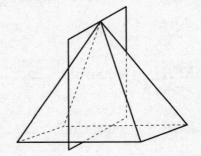

_____

Use the map below to answer questions 20 and 21. Use a ruler to measure the dimensions.

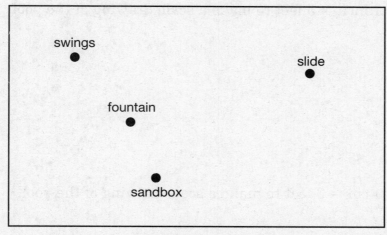

Scale: 1 cm = 3 m

20. What is the actual length of the playground? _____

21. What is the actual distance between the slide and the sandbox? _____

22. What is the area of the figure?

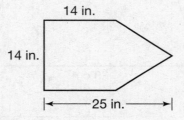

_____

Explain how you found your answer.

_____

_____

_____

23. The actual dimensions of a rectangular swimming pool are 30 feet long by 15 feet wide.

    Use a scale of 1 inch = 5 feet to make a scale drawing of the pool.

    Use a scale of 1 cm = 3 feet to make a scale drawing of the pool.

24. What is the area of the figure?

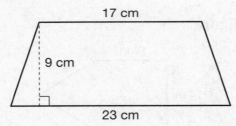

25. Draw the cross-section formed by the intersection of this figure with a plane that is parallel to its base.

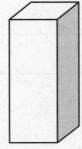

26. What is the surface area of the rectangular prism?

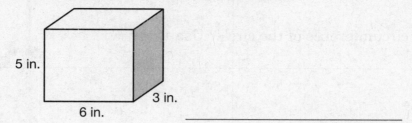

_____

27. An award is shaped like a prism whose base is a regular pentagon.

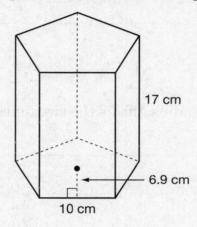

What is the volume of the award? _____

28. A figure consists of a rectangular prism and a cube.

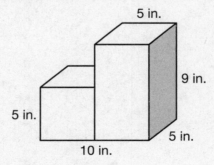

What is the surface area of the figure? _____

Explain how you found your answer.

_____

_____

29. The radius of a circle is 14 inches.

   Part A
   What is the circumference of the circle? Use $\frac{22}{7}$ for π.

   _____

   Part B
   Suppose the radius is doubled to 28 inches. What is the circumference of this circle?

   _____

   Part C
   Explain how doubling the radius affects the circumference.

   _____

   _____

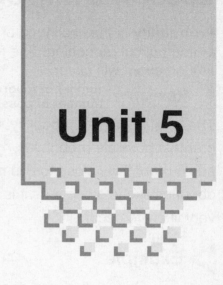

Unit 5

Statistics and Probability

Do you read news websites or subscribe to a magazine? If you do, you have probably seen examples of statistics. Journalists collect data and present it using tables, charts, and graphs. Many decisions and predictions are made based on data collected and analyzed from past observations. Statisticians (people who look for patterns in data) use terms such as mean, median, mode, and range when analyzing a data set. These interpretations are used in many types of news articles—whether about sports, or the economy, or politics. Probability has its roots in statistics, too. Through probability, we can make predictions based on the possible outcomes and how frequently those outcomes have occurred in the past. Weather forecasters use probability to determine the chance of rain or snow.

In this unit, you will calculate theoretical probabilities and compare them to experimental results. You will calculate the probability of compound events. You will learn about random and biased samples. You will find and use measures of center and measures of variation. You will find the mean absolute deviation of a data set. You will make predictions using data. Finally, you will compare data sets.

In This Unit

Probability

Compound Events

Samples

Measures of Central
   Tendency

Measures of Variation

Mean Absolute Deviation

Making Predictions Using
   Data

Comparing Data Sets

CCSS: 7.SP.5, 7.SP.6, 7.SP.7.a, 7.SP.7.b

Lesson 24: Probability

Probability is the likelihood of an event occurring. **Theoretical probability** is based on mathematical reasoning. The following formula can be used to find the probability, *P*, that an event will occur.

$$P\textbf{(event)} = \frac{\textbf{number of favorable outcomes}}{\textbf{number of possible outcomes}}$$

The least possible probability, 0, means that an event is impossible to occur. The greatest possible probability, 1, means that an event is certain to occur. An event that has a probability close to 0 is not likely to occur. An event that has a probability near $\frac{1}{2}$ is about as likely to occur as it is not to occur. An event that has a probability close to 1 is very likely to occur.

▷ **Example**

On which shape does the spinner have the least probability of landing? On which shape does the spinner have the greatest probability of landing?

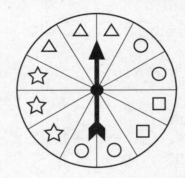

Find the probability of the spinner landing on each shape.

P (triangle) $= \frac{\text{number of favorable outcomes}}{\text{number of possible outcomes}} = \frac{3}{12} = \frac{1}{4}$

P (circle) $= \frac{\text{number of favorable outcomes}}{\text{number of possible outcomes}} = \frac{4}{12} = \frac{1}{3}$

P (square) $= \frac{\text{number of favorable outcomes}}{\text{number of possible outcomes}} = \frac{2}{12} = \frac{1}{6}$

P (star) $= \frac{\text{number of favorable outcomes}}{\text{number of possible outcomes}} = \frac{3}{12} = \frac{1}{4}$

The probability of the spinner landing on a square is $\frac{2}{12}$ or $\frac{1}{6}$. Because this is the least fraction, the spinner has the least probability of landing on a square. The probability of the spinner landing on a circle is $\frac{4}{12}$ or $\frac{1}{3}$. Because this is the greatest fraction, the spinner has the greatest probability of landing on a circle.

◆ **TIP:** When probability is expressed as a fraction, it is usually written in simplest form.

CCSS: 7.SP.5, 7.SP.6, 7.SP.7.a, 7.SP.7.b

Experimental probability is based on actual experiments and can be used to make predictions. Experimental probability is also called **relative frequency**.

The following formula can be used to find the experimental probability of an event.

$$\text{exp } P(\text{event}) = \frac{\text{number of times an event occurs}}{\text{number of trials}}$$

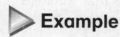 **Example**

David tossed a coin 15 times. The following table shows the outcomes. What is the experimental probability of getting tails?

Toss	1	2	3	4	5	6	7	8	9	10	11	12	13	14	15
Outcome	H	H	T	T	T	H	T	T	T	H	H	H	H	H	H

The number of times tails was tossed is 6. The number of trials performed is 15. Substitute these numbers into the formula above.

$$\text{exp } P(\text{tails}) = \frac{\text{number of actual outcomes}}{\text{number of trials}}$$
$$= \frac{6}{15}$$
$$= \frac{2}{5}$$

The experimental probability of getting tails is $\frac{2}{5}$.

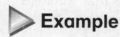 **Example**

David decides to toss the coin in the experiment above 100 more times. Based on the results of the experiment above, how many times should David expect the coin to land on tails over the next 100 tosses?

The experimental probability of the coin landing on tails is $\frac{2}{5}$.

Multiply the experimental probability by the number of tosses.

$$\frac{2}{5} \cdot 100 = \frac{200}{5} = 40$$

Based on the results of this experiment, David should expect the coin to land on tails 40 times in the next 100 tosses.

TIP: The experimental probability can change with every new trial you perform.

Experiments can be performed to test a theoretical probability. The theoretical probability of tossing tails on a coin is $\frac{1}{2}$. In the previous example, the experimental probability of tossing tails is $\frac{2}{5}$. As more trials are performed in an experiment, the experimental probability should get closer to the theoretical probability.

If you know the theoretical probability of an event and the number of trials that will take place, you can predict the number of times you can expect the event to occur.

 Example

About how many times would you expect the spinner to land on an odd number in 120 spins?

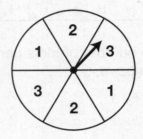

Find the theoretical probability of the spinner landing on an odd number.
$P(\text{odd}) = P(1 \text{ or } 3) = \frac{4}{6} = \frac{2}{3}$

Multiply the probability by the number of trials.
$\frac{2}{3} \cdot 120 = \frac{240}{3} = 80$

The spinner should land on an odd number about 80 times in 120 spins.

 Example

The following people are running for class president: Tim, Jane, Matthew, Linda, Maria, Joseph, Wendy, and Tina. Find the probability of a girl being selected as class president and the probability of Maria being selected as class president.

$P(\text{a girl is selected as class president}) = \dfrac{\text{number of girls running}}{\text{number of students running}} = \dfrac{5}{8}$

$P(\text{Maria is selected as class president}) = \dfrac{\text{Maria winning}}{\text{number of students running}} = \dfrac{1}{8}$

The probability that a girl is selected as class president is $\frac{5}{8}$ and the probability that Maria is selected as class president is $\frac{1}{8}$.

CCSS: 7.SP.5, 7.SP.6, 7.SP.7.a, 7.SP.7.b

▷ **Example**

A number cube numbered 1 through 6 is tossed and the spinner below is spun. What is the probability of getting a number greater than 3 on the cube? On the spinner? Are the outcomes equally likely?

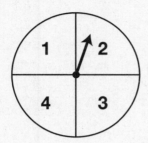

Find the probability of tossing a number greater than 3 on the number cube.

The numbers greater than 3 are 4, 5, and 6, so there are 3 favorable outcomes.
There are 6 possible outcomes.
P(greater than 3 on number cube) $= \frac{3}{6} = \frac{1}{2}$

Find the probability of spinning a number greater than 3.

The number 4 is the only number greater than 3 on the spinner, so there is 1 favorable outcome.
There are 4 possible outcomes.
P(greater than 3 on spinner) $= \frac{1}{4}$

The probability of getting a number greater than 3 on the number cube is $\frac{1}{2}$ and the probability of getting a number greater than 3 on the spinner is $\frac{1}{4}$. The outcomes are not equally likely because the number of possible outcomes is different for the number cube and the spinner.

CCSS: 7.SP.5, 7.SP.6, 7.SP.7.a, 7.SP.7.b

 Practice

Directions: Use the spinner below for questions 1 through 6. Find the probability of each event.

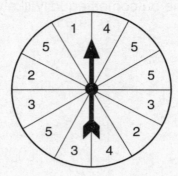

1. *P* (0) _____

2. *P* (odd number) _____

3. *P* (3) _____

4. *P* (2) _____

5. *P* (1) _____

6. *P* (number less than 6) _____

Directions: Use the following information for questions 7 through 12. Find the probability of each event.

Tanya has a box that holds 13 blue marbles, 9 purple marbles, 3 white marbles, and 20 red marbles. She pulls one marble out of the box without looking.

7. *P* (blue) _____

8. *P* (purple) _____

9. *P* (red) _____

10. *P* (white) _____

11. Is it more likely, less likely, or equally likely that Tanya will pick a blue marble than a purple marble from the box?

_____

12. Is it likely or unlikely that Tanya will pick a white marble from the box? Explain.

_____

_____

CCSS: 7.SP.5, 7.SP.6, 7.SP.7.a, 7.SP.7.b

13. If you flip a coin, what is the theoretical probability of the coin landing heads up?

 _____

14. If you flip a coin, what is the theoretical probability of the coin landing tails up?

 _____

15. If you flip a coin 40 times, predict the number of times it will land heads up and tails up.

 heads: _____   tails: _____

16. Flip a coin in the air 40 times and record your results in the following table.

Outcome	Number of Times
Heads (H)	
Tails (T)	

Directions: For questions 17 through 19, refer to your answers for questions 13 through 16.

17. What is your experimental probability of heads landing up? _____

18. What is your experimental probability of tails landing up? _____

19. How do your experimental probabilities compare to your predictions?

 _____

 _____

 _____

CCSS: 7.SP.5, 7.SP.6, 7.SP.7.a, 7.SP.7.b

Directions: Use the figures below to answer questions 20 through 23.

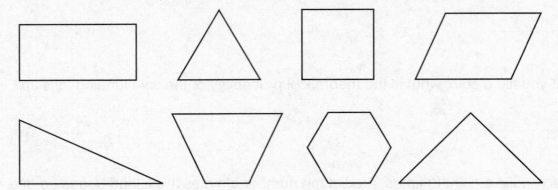

20. If one of the figures is selected at random, what is the probability that it is a triangle?

_____

21. If one of the figures is selected at random, what is the probability that it is a right triangle?

_____

22. If one of the figures is selected at random, what is the probability that it is a quadrilateral?

_____

23. If one of the quadrilaterals is selected at random, what is the probability that it has at least one right angle?

_____

24. If a number cube is rolled once, what is the probability of an even number facing upward?

    A. $\frac{1}{3}$

    B. $\frac{1}{2}$

    C. $\frac{2}{3}$

    D. $\frac{5}{6}$

25. If a number cube is rolled three hundred times, what is the expected number of times that number 2 or number 4 will be facing upward?

    A. about 10 times

    B. about 50 times

    C. about 60 times

    D. about 100 times

CCSS: 7.SP.5, 7.SP.6, 7.SP.7.a, 7.SP.7.b

26. Suppose you spin a spinner numbered 1 through 6, and you spin a 3 five times in a row. What is the theoretical probability of spinning a 3 on the next spin?

_____

Explain your answer.

_____

27. A paper cup in the shape of a cone is tossed. It can land either with its open end down or on its side, as shown.

Can the theoretical probability of the cup landing on its side be calculated based on the information given? Explain your answer.

_____

_____

Perform an experiment to determine the experimental probability of the paper cup landing on its side. Conduct at least 50 trials and record your results in the table.

Paper Cup Experiment

Number of times landed on side	
Number of times landed open end down	

What is the relative frequency of the cup landing on its side? _____

28. Mrs. Green made up quizzes for her math class. The quiz included one true-false question and one multiple-choice question with 4 possible answer choices.

What is the probability of getting the true-false question correct?_____

What is the probability of getting the multiple-choice question correct? _____

Are the outcomes equally likely? Explain your answer.

_____

_____

CCSS: 7.SP.8.a, 7.SP.8.b, 7.SP.8.c

Lesson 25: Compound Events

Compound events involve two or more simple events.

Two or more events that have no influence on each other are called **independent events**. To find the probability of independent events A and B, multiply the probability of the first event by the probability of the second event. Use the following formula.

$$P(A \text{ and } B) = P(A) \cdot P(B)$$

 Example

Alana has 6 blouses: 1 red, 1 blue, 2 striped, and 2 white. She has 4 skirts: 1 black, 1 blue, and 2 plaid. If Alana reaches into her closet without looking, what is the probability that she will pick a white blouse and a plaid skirt?

Step 1: Find the probability of picking a white blouse.

$P(\text{white blouse}) = \frac{2}{6} = \frac{1}{3}$

Step 2: Find the probability of picking a plaid skirt.

$P(\text{plaid skirt}) = \frac{2}{4} = \frac{1}{2}$

Step 3: Multiply the two probabilities.

$P(\text{white blouse and plaid skirt}) = P(\text{white blouse}) \cdot P(\text{plaid skirt})$

$= \frac{1}{3} \cdot \frac{1}{2}$

$= \frac{1}{6}$

The probability that Alana will pick a white blouse and plaid skirt is $\frac{1}{6}$.

CCSS: 7.SP.8.a, 7.SP.8.b, 7.SP.8.c

Two or more events that are influenced by each other are **dependent events**. To find the probability of dependent events A and B, multiply the probability of the first event by the probability of the second event. Use the following formula.

$$P(\text{A and B}) = P(\text{A}) \cdot P(\text{B})$$

 Example

> There are 6 lollipops in a bag: 4 are red and 2 are yellow. Two were picked at random, one after the other, without the first lollipop being replaced. What is the probability of picking a red lollipop first and a yellow lollipop second?

Step 1: Find the probability of the first event.
$$P(\text{red lollipop}) = \frac{4}{6} = \frac{2}{3}$$

Step 2: Find the probability of the second event.
Since there are only 5 lollipops left in the bag, the denominator is 5.
$$P(\text{yellow lollipop}) = \frac{2}{5}$$

Step 3: Multiply the two probabilities.
$$P(\text{red and yellow}) = P(\text{red lollipop}) \cdot P(\text{yellow lollipop})$$
$$= \frac{2}{3} \cdot \frac{2}{5}$$
$$= \frac{4}{15}$$

The probability of picking a red lollipop first and a yellow lollipop second is $\frac{4}{15}$.

You can use a list to show all possible outcomes.

 Example

> What is the probability of tossing two coins and getting heads on both coins?
>
> The following list shows the possible outcomes of tossing two coins.
> H, H   H, T   T, H   T, T
>
> The list shows that there are 4 possible outcomes. Only one possible outcome is 2 heads.
>
> The probability of getting heads on both coins is $\frac{1}{4}$.

A **tree diagram** uses branches to show all possible outcomes of an event.

 Example

Felix is going to buy a pair of shorts, a shirt, and a pair of shoes at the store. He wants to choose from two pairs of shorts (khaki or green), three shirts (white, yellow, or orange), and two pairs of shoes (sandals or sneakers). What is the probability of Felix choosing a pair of green shorts, a yellow shirt, and sandals?

The following tree diagram shows all the possible outcomes.

Shorts	Shirt	Shoes	Outcome
khaki	white	sandals	khaki shorts, white shirt, sandals
		sneakers	khaki shorts, white shirt, sneakers
	yellow	sandals	khaki shorts, yellow shirt, sandals
		sneakers	khaki shorts, yellow shirt, sneakers
	orange	sandals	khaki shorts, orange shirt, sandals
		sneakers	khaki shorts, orange shirt, sneakers
green	white	sandals	green shorts, white shirt, sandals
		sneakers	green shorts, white shirt, sneakers
	yellow	sandals	green shorts, yellow shirt, sandals
		sneakers	green shorts, yellow shirt, sneakers
	orange	sandals	green shorts, orange shirt, sandals
		sneakers	green shorts, orange shirt, sneakers

Felix can choose from 12 possible outcomes of a pair of shorts, a shirt, and a pair of shoes. Only one of these outcomes is green shorts, a yellow shirt, and sandals.

The probability of Felix choosing a pair of green shorts, a yellow shirt, and sandals is $\frac{1}{12}$.

CCSS: 7.SP.8.a, 7.SP.8.b, 7.SP.8.c

A **simulation** is a method of solving a problem by carrying out an experiment that is similar to the problem you need to solve.

A table of random digits can be used as a simulation tool to solve problems with compound events.

 Example

Use a table of random digits as a simulation tool to approximate an answer to the following situation: A machine correctly makes a paper cup 90% of the time. What is the probability that more than 3 out of 10 cups are defective?

Assign the digits 0 through 9 to outcomes for the situation. The digits 1 through 9 will correspond to non-defective cups. The digit 0 will correspond to a defective cup. Each trial consists of selecting ten digits. The number of defective cups in each trial is indicated by the number of 0s in the trial.

Sample of Random Digits

3809077894	1363784245	1272054317
0235837943	1040571252	9539159190
0704979042	9322926732	4266588239
0918020808	6183251350	4358060605
6570439876	4016565191	3825502149

Trial 1:   3809077894   There are 2 defective cups.

Trial 2:   1363784245   There are no defective cups.

Trial 3:   1272054317   There is 1 defective cup.

Trial 4:   0235837943   There is 1 defective cup.

Trial 5:   1040571252   There are 2 defective cups.

Trial 6:   9539159190   There is 1 defective cup.

Trial 7:   0704979042   There are 3 defective cups.

Trial 8:   9322926732   There are no defective cups.

Trial 9:   4266588239   There are no defective cups.

Trial 10:   0918020808   There are 4 defective cups.

Out of 10 trials, 1 trial had more than 3 out of 10 cups defective.

Based on the simulation, the probability that more than 3 out of 10 cups are defective is $\frac{1}{10}$ or 10%.

Practice

1. Claire is going to flip a coin three times. What is the probability of the coin landing tails up on the first flip and heads up on both of the last two flips?

   _____

2. Tim's golf bag contains 9 white golf balls, 6 yellow golf balls, and 1 orange golf ball. Without looking, Tim is going to pull one golf ball out of his bag to use for practice and then pull out a second golf ball for playing a round of golf. What is the probability of Tim practicing with a white ball and playing with an orange ball?

   _____

3. A drawer contains 10 blue pens, 10 black pens, and 10 red pens. Without looking, Mr. Lopez is going to take one pen from the drawer, use it, and then put it back into the drawer. Then he is going to take another pen from the drawer to use. What is the probability of Mr. Lopez taking a red pen first and then taking a blue pen?

   _____

4. There are 5 slices of pepperoni pizza, 1 slice of sausage pizza, and 3 slices of cheese pizza left at the pizza party. Without looking, Amy took a slice of pizza, ate it, and then took another slice. What is the probability of Amy eating two slices of cheese pizza?

   _____

5. Mrs. Cole has to choose two pieces of fruit for her lunch. If she randomly takes two pieces of fruit from a bowl of 8 oranges and 8 apples, what is the probability she will choose two apples? (Mrs. Cole does not put back the first piece of fruit before she takes the second piece of fruit.)

   A. $\frac{1}{4}$

   B. $\frac{7}{30}$

   C. $\frac{1}{32}$

   D. $\frac{1}{64}$

6. Tim, Jonah, and Chris are standing in line to buy tickets at the movies. Make a list to show the different ways they can stand in line.

_____

_____

How many different ways can they stand in line?

_____

7. Mandy is going to buy a new sweatshirt. She will choose the style (pull-over or zip-up), the size (L or XL), and the color (blue, yellow, or gray).

Draw a tree diagram to show all the choices Mandy has for a sweatshirt.

What is the probability of Mandy choosing a large zip-up blue sweatshirt?

_____

8. Two number cubes are rolled. Complete the table by listing the possible outcomes as ordered pairs.

First Number Cube

		1	2	3	4	5	6
Second Number Cube	**1**	(1, 1)	(2, 1)	(3, 1)	(4, 1)	(5, 1)	(6, 1)
	2						
	3						
	4						
	5						
	6						

What is the probability of rolling 4 twice?

_____

9. When playing baseball, Chad misses on 60% of the pitches he receives. Describe how to use random digits to simulate the probability that he will get a hit on more than 2 of every 10 pitches he receives.

_____

_____

_____

Use five trials of ten digits from the random digits table below to determine the probability that Chad will get a hit on more than 2 of every 10 pitches he receives. Explain.

Random Digits

4527372682	3831525034	3321635287
9766155286	3515453236	7663273974
9289266217	4579455151	9719527700
1030218610	5369436082	8031797247
2227257269	7925795725	0962442274

_____

_____

_____

CCSS: 7.SP.1, 7.SP.2

Lesson 26: Samples

You make generalizations or predictions about a large group called a **population** by examining a smaller group within that population called a **sample**. A sample is considered to be a **random sample** if each member of the population has an equal chance of being selected for it. A **representative sample** is useful because it has characteristics that are similar to those of the larger population.

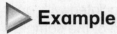 **Example**

Jack wants to sample students to find out how many books they read over the summer. Which of the following groups would be a random sample?
A. all the students in the library
B. all the students in an honors math class
C. every tenth student entering the lunchroom

The students in the library do not represent the total student population because not every student would visit the library.

The students in the honors math class do not represent the total student population because not every student could be in the honors math class.

Every tenth student entering the lunchroom represents the total student population because every student has to enter the lunchroom.

Choice C would be a random sample.

A random sample is **unbiased** since every individual in the population has an equal chance of being selected. When the sample size is large and random, the data gathered from that sample will better represent the entire population. A **biased sample** is less random and favors certain characteristics. The data gathered from a biased sample is usually not reliable for making generalizations about the entire population.

▶ Example

You want to find out what type of vehicle is most popular among drivers in your town. To create a sample, you decide to walk around a school parking lot and survey people who drive SUVs as they wait to pick up their children. Is this sample biased? If so, how can you make the sample less biased?

The sample you created is biased for several reasons. First, it only includes drivers of one type of vehicle. Second, it only includes people associated with one school, which may reduce the social, cultural, and economic range of the sample. To make the sample less biased, you can have a person stand outside the main entrances of shopping centers or large public parking lots and survey every tenth adult who says he or she drives a motorized vehicle. The sample is now less biased since there is a better chance of asking drivers of vehicles other than SUVs and drivers from a diverse set of backgrounds.

● Practice

Directions: For questions 1 through 4, explain whether or not the situation would provide a random sample.

1. determining the favorite candidate for seventh-grade class president by using the results of a voluntary online survey

   _____

   _____

2. determining the nation's favorite ice cream flavor by asking a group of customers in one ice cream shop

   _____

   _____

3. determining how guests rate the food at a certain restaurant by asking every tenth person who leaves the restaurant

   _____

   _____

4. determining the fruit preferred by most seventh graders at one school by asking every tenth student entering each seventh-grade classroom

   _____

   _____

Directions: For questions 5 through 7, identify whether the situation would provide a random or biased sample.

5. determining whether the Springfield Tigers or the Washington Lions are the better sports team by asking citizens of Springfield

_____

6. determining the average yearly rainfall in Kentucky by measuring the rainfall in various counties in northern, central, and southern Kentucky

_____

7. determining which sport is the most popular among adults by surveying every fifth person who arrives to watch a tennis match

_____

8. Which situation would provide a random sample?

   A. determining the most popular choice for a career in the country by surveying every tenth student at one college campus

   B. determining the favorite vegetable of seventh graders in your school district by surveying every twentieth student entering each seventh-grade classroom in your district

   C. determining the average income in your state by surveying residents of the three richest communities in your state

   D. determining whether voters are enthusiastic about an election by surveying people in line to vote

9. You want to find out which type of cereal is most popular among children in your town. You decide to survey customers at a grocery store who are buying cereal.

   Explain why the sample is biased.

   _____

   _____

   Explain how you can make the sample less biased.

   _____

   _____

Lesson 27: Measures of Central Tendency

The **mean**, **median**, and **mode** are **measures of central tendency**. A measure of central tendency is a single number used to represent all the values in a data set.

Mean

The mean is the sum of the values in a data set divided by the number of values in the set. It is affected by all the numbers in the set.

Median

The median is the middle number in a data set that has been arranged in order of value. If the number of data values is odd, the median is the middle value. If the number of data values is even, the median is the average of the two middle values.

Mode

The mode is the number that appears most often in a data set. A data set may contain one mode, more than one mode (if two or more values appear most often), or no mode at all (if no value appears more than once in the set).

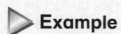 **Example**

Find the mean, median, and mode of the following data set.
70, 71, 71, 71, 75, 77, 82, 83, 83, 87

To find the mean, add the values and divide by the number of addends.

$$\text{mean} = \frac{70 + 71 + 71 + 71 + 75 + 77 + 82 + 83 + 83 + 87}{10} = \frac{770}{10} = 77$$

There is an even number of data values. The two middle values are 75 and 77.

$$\text{median} = \frac{75 + 77}{2} = \frac{152}{2} = 76$$

The number 71 occurs 3 times.
mode = 71

The mean is 77, the median is 76, and the mode is 71.

 TIP: The most familiar measure of central tendency is the mean (average), but sometimes other measures of central tendency provide a more accurate representation of the data.

CCSS: 7.SP.4

Although the mean is usually the most accurate measure of central tendency for describing a data set, it is sometimes better to use the median or the mode.

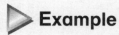 **Example**

The following table shows the number of tickets sold over one week for a play.

Day	Number of Tickets Sold
Sunday	230
Monday	175
Tuesday	160
Wednesday	180
Thursday	210
Friday	230
Saturday	460

Find the mean number of tickets sold for the week.
$$\text{mean} = \frac{230 + 175 + 160 + 180 + 210 + 230 + 460}{7} = \frac{1645}{7} = 235$$

Find the mode.
Since the number 230 occurs twice, it is the mode.

Find the median.
Arrange the numbers in order from least to greatest.
160, 175, 180, 210, 230, 230, 460
The median, the middle number, is 210.

The mean number of tickets sold, 235, is higher than six of the seven numbers. The mean is not representative of this data set, because its value is influenced by one extremely high value, 460.

The mean is usually the measure of central tendency that best represents a data set. However, if a data set includes an **outlier**, the median or mode will most likely be a better choice. An outlier is a value that is noticeably smaller or larger than the rest of the data values. In this example, the median number of tickets sold, 210, is a more accurate measure of central tendency.

Sometimes two measures of central tendency can represent a data set better than any single measure.

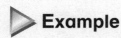 **Example**

Emilio manages a restaurant. He is getting ready to order more supplies. The list below shows the number of customers who visited Emilio's restaurant each day last week.

88, 220, 253, 305, 310, 310, 306
mean: 256   median: 305   mode: 310

Emilio needs to use the best measure of central tendency to predict how many customers will visit the restaurant next week. Which measure of central tendency (mean, median, or mode) is best for him to use?

The mean (256) represents the average number of customers Emilio had each day. If Emilio uses the mean when ordering supplies, he may not order enough. Emilio had at least 256 customers on four of the seven days last week. If Emilio assumes an average of 256 customers each day, he will probably run out of supplies. In this case, the mean has been lowered by an outlier. The number 88 is an outlier, which makes the mean not very accurate in describing this data set.

The median (305) represents the middle value of Emilio's data set and gives a more accurate representation of the number of customers Emilio will have each day.
The mode (310) represents the number of customers that Emilio had "most often" in one day. It also seems to be an accurate representation of the number of customers Emilio will have each day.

Therefore, either the median or the mode is the best measure for Emilio to use when ordering supplies.

CCSS: 7.SP.4

A **dot plot** is often used to display data. A dot plot uses a number line and dots to indicate data values. The number of dots above each value shows how many times that value occurs in the data set. A dot plot shows the spread of the numbers in a data set. It allows you to quickly identify the mode, any **clusters** of data values, and any outliers.

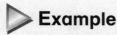 **Example**

The following list shows the number of points scored by a football team in the 16 games it played this season.
17, 7, 30, 28, 30, 17, 20, 10, 27, 21, 6, 19, 17, 10, 23, 17

The data from the list are displayed in the following dot plot.

Points Scored in Football Games

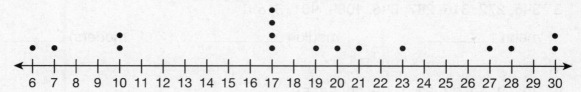

The dot plot shows the spread of the data values in an organized way. It also allows you to see the data differently than you would with a list. From the dot plot, you can quickly see that the mode is 17.

You can also see that the data is clustered between 17 and 23.

You can find the median number of points scored by finding the middle dot. Since there is an even number of scores, the two middle scores are 17 and 19. The median is $\frac{17 + 19}{2} = \frac{36}{2} = 18$.

⬤ Practice

Directions: For questions 1 through 5, find the mean, median, and mode(s) of each data set. Round your answer to the nearest tenth when necessary.

1. 35, 104, 57, 92, 29, 46, 57

   mean _____   median _____   mode(s) _____

2. 63, 136, 187, 47, 106, 33, 213, 111

   mean _____   median _____   mode(s) _____

3. 346, 272, 316, 287, 346, 1084, 401, 316

   mean _____   median _____   mode(s) _____

4. 70, 53, 42, 23, 53, 1, 45, 29, 28, 56

   mean _____   median _____   mode(s) _____

5. 125, 216, 162, 347, 862, 254, 119, 203

   mean _____   median _____   mode(s) _____

Directions: Use the following information to answer questions 6 through 9.

The following list shows the scores that Alicia received on her math tests.

94   86   92   100   66   99   91   99   83

6. Find the mean, median, and mode of this data set.

   mean: _____   median: _____   mode: _____

7. Which measure(s) of central tendency would be appropriate to describe Alicia's math test scores? Explain your answer.

   _____

   _____

8. The teacher decides not to count the lowest score. Find the mean, median, and mode of the remaining 8 scores.

   mean: _____   median: _____   mode: _____

9. Which measure of central tendency is affected most by a very low score? Explain your answer.

   _____

Directions: The table shows the number of points each player scored in 7 basketball games. Use the table to answer questions 10 through 12.

Points Scored in Basketball Games

	Game 1	Game 2	Game 3	Game 4	Game 5	Game 6	Game 7
Lance	7	10	5	7	12	8	7
Hannah	11	13	12	16	7	15	10
Randy	8	9	14	26	10	11	6

10. Which measure(s) of central tendency would be appropriate to describe the number of points Lance scored each game?

    _____

    Explain your answer.

    _____

    _____

11. Which measure(s) of central tendency would be appropriate to describe the number of points Hannah scored each game?

    _____

    Explain your answer.

    _____

12. Which measure(s) of central tendency would be appropriate to describe the number of points Randy scored each game?

    _____

    Explain your answer.

    _____

13. At the vet's office where Terry works, she was asked to find the median and mode weights of a litter of puppies. The following list shows the weights of the puppies, in ounces.
    18, 24, 20, 16, 14, 27, 24

    What are the median and mode weights?

    A. Both measures are 24.

    B. The median is 20 and the mode is 24.

    C. Both measures are 20.

    D. The median is 24 and the mode is 20.

14. Why should Eric be careful using the mean for the following data set?
    23, 30, 51, 19, 32, 41, 38, 25, 48, 41, 8

    A. The mean is not a whole number.

    B. The data set has no mode.

    C. The data set has an outlier of 8 which will affect the mean.

    D. The data set has an odd number of values.

15. The following dot plot shows the daily high temperature, in °F, for each day in October.

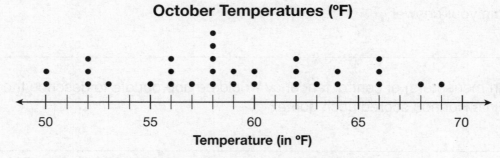

October Temperatures (°F)

Temperature (in °F)

What was the mode of the high temperatures for October?

_____

What was the median high temperature for October?

_____

Explain how you found the median high temperature.

_____

_____

_____

Lesson 28: Measures of Variation

A **measure of variation** shows how values are spread out within a data set. For example, the **range** of a data set is the difference between the largest and smallest values in the data set.

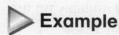 Example

What is the range of the following data set?
103, 109, 85, 141, 168, 117, 128, 102, 98

The smallest number is 85 and the largest number is 168.
Subtract the smallest number from the largest number.
$168 - 85 = 83$

The range of the data set is 83.

A **quartile** is one fourth of the values in a data set. The **lower quartile (Q1)** is the median of the lower half of the data set. The **second quartile (Q2)** is the median of the entire data set. The **upper quartile (Q3)** is the median of the upper half of the data set. The median of the data set is not included in the values used to find the lower and upper quartiles. The **interquartile range (IQR)** of a data set is the difference between the values of the upper quartile and the lower quartile.

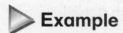 Example

What are the lower quartile, upper quartile, and interquartile range of the following data set?
36, 17, 5, 22, 36, 43, 13, 27, 40

Arrange the numbers from least to greatest, and find the median. Then find the median of the lower half and the median of the upper half.

$$\boxed{5, 13, 17, 22,} \quad 27, \quad \boxed{36, 36, 40, 43}$$

$$\begin{array}{ccc} \uparrow & \uparrow & \uparrow \\ 15 & 27 & 38 \\ Q1 & \text{median} & Q3 \end{array}$$

The lower quartile is $\frac{13 + 17}{2}$, or 15, and the upper quartile is $\frac{36 + 40}{2}$, or 38.

Subtract the lower quartile from the upper quartile.
$38 - 15 = 23$

The interquartile range is 23.

A **percentile** is a measure that indicates what percent of the total number of values in the data set is scored at or below that measure. The pth percentile of a data set is a value for which at least p percent of the items have this value or less. This also means that at least $(100 - p)$ percent of the items have this value or more.

To find the pth percentile of a data set, arrange the data in order from least to greatest. Find the **index (i)**, the position of the pth percentile in the ordered set, by multiplying the percent (written as a decimal) by the total number of values in the data set. If the product is not an integer, round up. Then find the data value in the ith position.

 Example

> The following list represents the number of spelling words that the 15 students in Kiran's class got correct on the last spelling test.
> 12, 15, 16, 17, 18, 19, 19, 20, 21, 22, 23, 23, 24, 24, 25
>
> If the number of spelling words that Kiran got correct was at the 85th percentile, how many spelling words did she get correct?
>
> Convert 85% to a decimal (0.85) and multiply by the total number of values in the data set.
> $0.85 \cdot 15 = 12.75$
>
> Since 12.75 is not an integer, round up to 13. The number of spelling words that Kiran got correct is the 13th number listed. Therefore, Kiran got 24 spelling words correct.

If the product is an integer, the pth percentile is the average of the values in positions i and $i + 1$.

 Example

> Using the same set of numbers in the example above, what value represents the 60th percentile?
>
> Convert 60% to a decimal (0.6) and multiply by the total number of values in the data set.
> $0.6 \cdot 15 = 9$
>
> Since 9 is an integer, the 60th percentile is the average of the 9th and 10th scores. These values are 21 and 22. The 60th percentile is 21.5.

 TIP: The 25th percentile is the lower quartile (Q1), the 50th percentile is the median, and the 75th percentile is the upper quartile (Q3) of the data set.

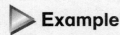
CCSS: 7.SP.4

A **box-and-whisker plot** shows how data is distributed, or spread out, along a number line. The box shows how the middle 50% of the data is grouped together, and the whiskers show how far away the minimum and maximum values are from the middle 50%. The plot also shows the median and the upper and lower quartiles of the data set.

▶ Example

The following list shows the scores of the first 9 games for a high school basketball team.
67, 59, 62, 64, 71, 71, 52, 53, 61

A box-and-whisker plot can be drawn to represent the data. Arrange the data values in order from least to greatest. Find the minimum, maximum, median, and quartiles.
52, 53, 59, 61, 62, 64, 67, 71, 71

The following diagram shows how to arrange and separate the data.

$$\boxed{52,\ 53,\ 59,\ 61}\ \ 62,\ \boxed{64,\ 67,\ 71,\ 71}$$

Use the diagram above to find the following values to include in the box-and-whisker plot.

The minimum value is 52.

The lower quartile is $53 + \frac{59}{2}$ or 56.

The median is 62.

The upper quartile is $67 + \frac{71}{2}$ or 69.

The maximum value is 71.

The following box-and-whisker plot displays the data.

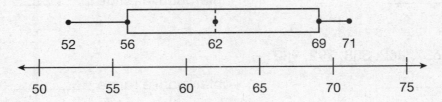

The box-and-whisker plot shows that 50% of the data has values from 56 to 71.

 TIP: You can draw conclusions about a data set by looking at the length of the box in the box-and-whisker plot. The shorter the box, the better the median is at describing the whole data set.

◯ Practice

Directions: Use the following information to answer questions 1 through 5.

The following table shows the scores that Deanna received on the last seven computer games she played.

Game Scores

Game	1	2	3	4	5	6	7
Score	53	59	34	48	63	55	63

1. What is the range of Deanna's scores? _____

2. What is the lower quartile of Deanna's scores? _____

3. What is the upper quartile of Deanna's scores? _____

4. What is the interquartile range of Deanna's scores? _____

5. Which score is the 60th percentile? _____

Directions: For questions 6 through 8, find the range and the interquartile range of the data set.

6. 31, 39, 29, 34, 37, 24, 36, 33, 34

   range _____   interquartile range _____

7. 278, 739, 637, 462, 308, 792, 942

   range _____   interquartile range _____

8. 44, 38, 13, 70, 44, 22, 66, 104, 83

   range _____   interquartile range _____

CCSS: 7.SP.4

Directions: Use the following table to answer questions 9 and 10.

Day	Sunday	Monday	Tuesday	Wednesday	Thursday	Friday	Saturday
Low Temperature	18°F	22°F	16°F	20°F	28°F	30°F	22°F

9. What is the interquartile range of the low temperatures for the week?

   A. 28°F   C. 14°F

   B. 18°F   D. 10°F

10. What temperature is at the 30th percentile?

   A. 18°F   C. 22°F

   B. 20°F   D. 30°F

11. The following list shows the annual number of wins that a baseball pitcher had in each of his twenty years as a baseball pitcher.

   9, 7, 24, 20, 18, 17, 21, 18, 18, 12, 9, 10, 10, 21, 20, 14, 13, 20, 13, 17

   What is the lower quartile? _____

   What is the median number of wins? _____

   What is the upper quartile? _____

   What is the maximum number of wins? _____

   Display the data in a box-and-whisker plot on the number line below.

   Number of Wins

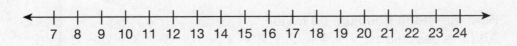

CCSS: 7.SP.3

Lesson 29: Mean Absolute Deviation

The **mean absolute deviation** is another measure of variation of a data set. It indicates how the data varies, or differs, from the mean.

To find the mean absolute deviation:

1. Find the mean of the data set.
2. Find the amount that each data point deviates from the mean by calculating the difference of the data point and the mean. Then find the absolute value of each difference.
3. Find the sum of the absolute values.
4. Divide the sum by the total number of values in the data set.

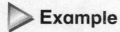 **Example**

What is the mean absolute deviation of the following data set?
9, 21, 24, 16, 17, 19, 11, 21, 26, 16

Find the mean.
9 + 21 + 24 + 16 + 17 + 19 + 11 + 21 + 26 + 16 = 180
180 ÷ 10 = 18

Make a table. List the numbers in order in the first column. Find the deviation from the mean in the second column. List the distance from the mean in the third column.

Number	Deviation from Mean	Distance from Mean
9	9 − 18 = −9	9
11	11 − 18 = −7	7
16	16 − 18 = −2	2
16	16 − 18 = −2	2
17	17 − 18 = −1	1
19	19 − 18 = 1	1
21	21 − 18 = 3	3
21	21 − 18 = 3	3
24	24 − 18 = 6	6
26	26 − 18 = 8	8

Find the mean absolute deviation.
Add the deviations: 9 + 7 + 2 + 2 + 1 + 1 + 3 + 3 + 6 + 8 = 42
Divide by 10, the number of values in the data set.
mean absolute deviation = 42 ÷ 10 = 4.2

The mean absolute deviation is 4.2.

CCSS: 7.SP.3

When the mean absolute deviation is small, it means the data are clustered closely together. If the mean absolute deviation is large, it means the data are more spread out.

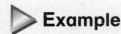 **Example**

Wendy bought some gifts for her family. The gifts cost $67, $28, $15, $79, $75, and $12. Display the data on a dot plot. Then find the mean absolute deviation of the cost of the gifts.

The dot plot shows the cost of the gifts.

Cost of Gifts (in dollars)

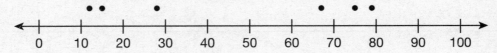

Notice that the data are spread out.

Find the mean.
67 + 28 + 15 + 79 + 75 + 12 = 276
276 ÷ 6 = 46

Make a table. List the costs in order in the first column. Find the deviation from the mean in the second column. List the distance from the mean in the third column.

Number	Deviation from Mean	Distance from Mean
12	12 − 46 = −34	34
15	15 − 46 = −31	31
28	28 − 46 = −18	18
67	67 − 46 = 21	21
75	75 − 46 = 29	29
79	79 − 46 = 33	33

Find the mean absolute deviation.
Add the deviations: 34 + 31 + 18 + 21 + 29 + 33 = 166

Divide this sum by 6, the number of values in the data set. Round the quotient to the nearest tenth.
mean absolute deviation = 166 ÷ 6 = 27.7

The mean absolute deviation is 27.7.

The mean absolute deviation is large, which indicates that the data are spread out, as shown on the dot plot.

Practice

Directions: For questions 1 through 4, find the mean absolute deviation of the data set. Round your answer to the nearest tenth when necessary.

1. 2, 5, 7, 9, 1, 3, 4, 2, 10, 7

_____

2. 63, 36, 18, 47, 19, 33

_____

3. 46, 72, 31, 28, 34, 13, 40, 16

_____

4. 20, 35, 24, 23, 13, 17, 15, 29, 28, 26

_____

5. Which data set has a mean absolute deviation of 9.2?

   A. 90, 75, 85, 100, 80

   B. 100, 90, 65, 90, 100

   C. 70, 80, 95, 100, 85

   D. 95, 100, 75, 80, 75

6. The golf scores of the members of a golf team for 18 holes were 89, 90, 87, 95, 86, 102, 79, and 108. What is the mean absolute deviation of the scores?

   A. 7.25

   B. 9.2

   C. 72.5

   D. 92

7. The time in minutes that it took five students to complete a math test was 35, 27, 30, 25, and 38.

   What is the mean absolute deviation of the times?

   _____

   Does the value of the mean absolute deviation indicate that the data were spread out or close together? Explain your answer.

   _____

   _____

8. Last week Jeremy kept a log of the number of miles that he biked.

Day	Number of Miles Biked
Sunday	15
Monday	6
Tuesday	5
Wednesday	0
Thursday	2
Friday	8
Saturday	13

   Use the number line below to display the data on a dot plot.

   Number of Miles Biked

   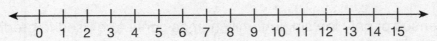

   Does the dot plot show that the data are spread out or close together?

   _____

   Find the mean absolute deviation of the number of miles biked to the nearest tenth.

   _____

   How does the value of the mean absolute deviation compare to your answer about the dot plot? Explain your answer.

   _____

   _____

223

CCSS: 7.SP.2, 7.SP.4

Lesson 30: Making Predictions Using Data

Sometimes data can be collected from a complete population. A **census** is a survey of the entire population. This data gives an excellent description of the population, but it is possible only if the population is not too large and is available to be surveyed. When the population is too large, data collected from a sample is used to make predictions about the entire population.

A sample may give very accurate information about a population. Or, you may realize a sample is biased. In that case, you can analyze the biased sample to determine how it relates to the population. For example, if you were trying to determine the average weight of a car, but the only available sample was taken from the largest-sized cars manufactured, you would expect the average weight of a car to be less than the average weight of the sample.

 Example

The town of Milton used a sample to determine how many people live in the entire town. A map of the area was used to divide the town into small sections called clusters. Some areas of Milton have apartment buildings and other areas have single-family homes. A cluster is selected at random and a census is completed on the people who live there. If the cluster selected consists of single-family homes only, can the survey be used to predict the entire population of Milton?

The number of people per square mile in an area consisting of apartment buildings is usually greater than the number of people per square mile in an area of single-family homes. Therefore, basing the census only on a sample of the single-family homes will underestimate the entire population of Milton.

A census of only the single-family homes cannot be used to predict the entire population of Milton.

Example

Using the example above, how do you think the population predicted by the sample would compare to the actual population of Milton?

The population predicted by the sample would be much less than the actual population of Milton.

CCSS: 7.SP.2, 7.SP.4

 Practice

Directions: For questions 1 through 3, evaluate whether the given measure of central tendency or variation from the sample can or cannot be applied to the entire population. If it cannot be applied to the entire population, describe how the measure would differ between the sample and the population.

1. Population: the median home value of a city
   Sample: the home values taken from the neighborhood in the city with the largest homes

   _____

   _____

   _____

2. Population: the range of ages of the entire population of a city
   Sample: the range of ages taken from a census of a few randomly selected sections of the city

   _____

   _____

   _____

3. Population: the mean salary of employees of a company
   Sample: the mean salaries of temporary employees at the company

   _____

   _____

   _____

4. Keeshon is collecting data about the favorite sports of the students in his class. Which of the following is an example of a census?

   A. Keeshon interviews the boys in his class.

   B. Keeshon interviews every third student to enter his classroom.

   C. Keeshon interviews every student in his class.

   D. Keeshon interviews every student at a basketball game.

5. The mayor of a city wants to find out what local voters think about the proposed expansion of a highway. To find out, she surveys the employees of a company that will work on the highway expansion. The results show that 90% of the employees favor the expansion.

   Can the mayor conclude that 90% of the residents of the city favor the expansion? Explain your answer.

   _____

   _____

   _____

6. Adults leaving a library were surveyed about the possibility of an annual fee of $100 to support the library. Of those surveyed, 80% said they were in favor of the fee. The survey takers concluded incorrectly that 80% of the adult residents of the city are in favor of the fee.

   What would be a better survey? What might be a more believable value for the percent of adult residents of the city who are in favor of the fee?

   _____

   _____

   _____

7. An auto repair shop wants to measure customer satisfaction with its work. It plans to conduct a survey of a random sample of repeat customers.

   Is this a random sample of all of the shop's customers? If not, explain how the sample is biased.

   _____

   _____

   _____

   What would be a better method of surveying customer satisfaction?

   _____

   _____

   _____

CCSS: 7.SP.3, 7.SP.4

Lesson 31: Comparing Data Sets

You can compare the measures of central tendency for two sets of data.

 Example

Mr. Barry finished scoring his students' math tests. The following stem-and-leaf plots display the data for the test scores for the girls and for the boys. Recall that each value is displayed in stem-and-leaf plots by separating digits. Here, the tens digits are the stems and the ones digits are the leaves, so the second row, 7 | 0 2 6 6, lists four data values: 70, 72, 76, and 76.

**Boys' Math
Test Scores**

Stem	Leaf
6	5
7	0 2 6 6
8	3 5 7
9	1 4 8 9
10	

Key: 7 | 0 = 70

**Girls' Math
Test Scores**

Stem	Leaf
6	8
7	4
8	5 6 8 9
9	1 5 5 7
10	0

Key: 8 | 5 = 85

Compare the mean, median, and mode for the girls' and boys' test scores.

Boys
Mean: 65 + 70 + 72 + 76 + 76 + 83 + 85 + 87 + 91 + 94 + 98 + 99 = 996
996 ÷ 12 = 83
Median: There are 12 scores for the boys, so the median score is halfway between the 6th and 7th scores.
$\frac{83 + 85}{2} = \frac{168}{2} = 84$
Mode: 76 appears most often, so the mode is 76.

Girls
Mean: 68 + 74 + 85 + 86 + 88 + 89 + 91 + 95 + 95 + 97 + 100 = 968
968 ÷ 11 = 88
Median: There are 11 scores for the girls, so the median score is the 6th score, 89.
Mode: 95 appears most often, so the mode is 95.

The mean test score for the girls is 5 points more than for the boys. The median test score for the girls is 5 points higher than for the boys. The mode for the boys is 76 and the mode for the girls is 95. By looking at the stem-and-leaf plots, you can see that the test scores for the girls are generally higher than the test scores for the boys.

▶ Example

Garrett and Ella drive similar cars. They calculate their gas mileage (how far they can travel in miles per gallon of gas) each time they fill up their gas tanks. The following dot plots display their gas mileage data over each of the last eight weeks.

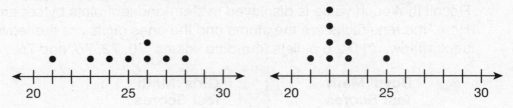

Gas Mileage for Garrett's Car   Gas Mileage for Ella's Car

Compare the mean, median, and mode gas mileage for Garrett's and Ella's cars.

Garrett
Mean: 26 + 28 + 21 + 24 + 26 + 23 + 25 + 27 = 200
200 ÷ 8 = 25
Median: There are 8 dots, so the median score is halfway between the 4th and 5th dot.
$\frac{25 + 26}{2} = \frac{51}{2} = 25.5$
Mode: 26 appears most often, so the mode is 26.

Ella
Mean: 22 + 23 + 21 + 22 + 25 + 22 + 23 + 22 = 180
180 ÷ 8 = 22.5
Median: There are 8 dots, so the median score is halfway between the 4th and 5th dot.
The 4th and 5th dots are both at 22, so the median is 22.
Mode: 22 appears most often, so the mode is 22.

The mean, median, and mode gas mileage is greater for Garrett's car than for Ella's car. By looking at the dot plots, you can see that the gas mileage for Garrett's car is usually higher than the gas mileage for Ella's car. You can also see that the data for Garrett's car is more spread out than the data for Ella's car. This larger spread of data for Garrett's car indicates that its gas mileage is less consistent than that of Ella's car.

CCSS: 7.SP.3, 7.SP.4

You can also use the mean absolute deviation to compare two data sets.

▷ Example

The following dot plots display the number of goals scored by the Jaguars and the Panthers soccer teams in each of their first ten games.

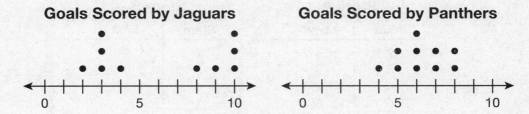

Goals Scored by Jaguars   **Goals Scored by Panthers**

Compare the mean and the mean absolute deviation of the two teams' data.

Jaguars
Mean: $2 + 3 + 3 + 3 + 4 + 8 + 9 + 10 + 10 + 10 = 62$
$62 \div 10 = 6.2$
Mean Absolute Deviation:
$4.2 + 3.2 + 3.2 + 3.2 + 2.2 + 1.8 + 2.8 + 3.8 + 3.8 + 3.8 = 32$
$32 \div 10 = 3.2$

Panthers
Mean: $4 + 5 + 5 + 6 + 6 + 6 + 7 + 7 + 8 + 8 = 62$
$62 \div 10 = 6.2$
Mean Absolute Deviation:
$2.2 + 1.2 + 1.2 + 0.2 + 0.2 + 0.2 + 0.8 + 0.8 + 1.8 + 1.8 = 10.4$
$10.4 \div 10 = 1.04$

The mean for both the Jaguars and the Panthers is 6.2. The mean absolute deviation for the Jaguars is about 3 times the mean absolute deviation for the Panthers, which means the Jaguars have a wider spread than the Panthers in the number of goals they scored. By looking at the dot plots, you can see that the number of goals that the Jaguars scored is more spread out than the number of goals that the Panthers scored.

Practice

Directions: Use the stem-and-leaf plots and information below to answer questions 1 through 3.

These plots show the ages of the employees in two stores owned by the same company.

Store A			Store B	
Stem	**Leaf**		**Stem**	**Leaf**
2	3 4 8		2	1 5
3	7 9		3	0 0 2 6 7 9
4	0 7 7		4	1 4
5	1 6			

Key: 3 | 7 = 37   **Key:** 3 | 0 = 30

1. Find the mean, median, and mode ages of the employees in Store A.

    mean: _____   median: _____   mode: _____

2. Find the mean, median, and mode ages of the employees in Store B.

    mean: _____   median: _____   mode: _____

3. Write four statements that compare the data of the ages of the employees in the two stores.

    _____

    _____

    _____

    _____

Directions: Use the dot plots to answer questions 4 through 7.

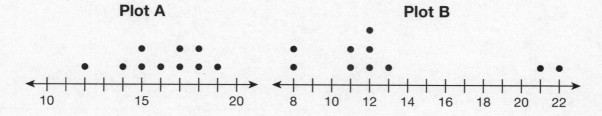

CCSS: 7.SP.3, 7.SP.4

4. Find the mean of the data in each dot plot.

   Plot A: _____   Plot B: _____

5. Which plot has the greater mean? How much greater? _____

6. Find the mean absolute deviation of the data in each dot plot.

   Plot A: _____   Plot B: _____

7. Which plot has the greater mean absolute deviation? About how many times greater is it?

   _____

8. The dot plots below show the hourly pay in dollars of 8 people with college degrees and 8 people without college degrees who work for the same company.

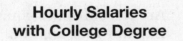

**Hourly Salaries
with College Degree**

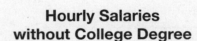

**Hourly Salaries
without College Degree**

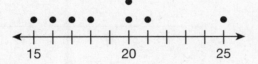

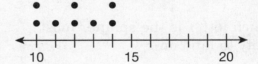

   Find the mean and the mean absolute deviation of the hourly pay of people with college degrees.

   mean: _____   mean absolute deviation: _____

   Find the mean and the mean absolute deviation of the hourly pay of people without college degrees.

   mean: _____   mean absolute deviation: _____

   How much greater is the mean hourly pay for those with college degrees than for those without college degrees?

   _____

   About how many times greater is the mean absolute deviation in hourly pay for those with college degrees than for those without college degrees?

   _____

Unit 5 Practice Test

Directions: Use the spinner to answer questions 1 through 6.

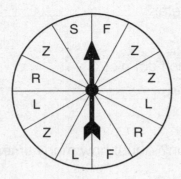

1. P(R) _____

2. P(L) _____

3. P(B) _____

4. Which letter is the spinner most likely to land on?

   _____

5. Which letter is the spinner least likely to land on?

   _____

6. If the spinner is spun 200 times, about how many times would you expect the spinner to land on an F or S?

   _____

7. Grace tossed a coin and rolled a number cube numbered 1 through 6.

   What is the probability of tossing heads on the coin?

   _____

   What is the probability of rolling an odd number on the number cube?

   _____

   Are the outcomes equally likely? Explain your answer.

   _____

   _____

   _____

8. Jermaine has 10 shirts in his drawer: 3 red, 2 black, and 5 white. He has 6 pairs of shorts in his drawer: 3 black, 1 blue, and 2 khaki. What is the probability that a shirt and a pair of shorts chosen at random are both black?

_____

9. Two number cubes are rolled and their sum is found. Complete the table for the possible outcomes.

First Number Cube

Second Number Cube		1	2	3	4	5	6
	1	2	3	4	5	6	7
	2						
	3						
	4						
	5						
	6						

What is the probability of rolling two number cubes and getting a sum of 6?

_____

10. Use the random digits table to answer the questions below.

7309188708	6841329297	2658823756
1795227565	9872685244	9913057473
5941384972	3394849025	2436108339
5649512287	7135940787	3250110475
4999631433	1368040383	6027327161

While testing light bulbs, the manufacturer finds a defective light bulb 20% of the time.

Describe how to use the random digits table in ten trials of a simulation to find the probability that more than 2 out of 10 light bulbs are defective.

_____

_____

_____

What is the probability that more than 2 out of 10 light bulbs are defective?

_____

11. A bowl contains 6 red candies, 8 orange candies, and 10 yellow candies. What is the probability that a candy chosen at random will be yellow?

    A. $\frac{3}{10}$

    B. $\frac{3}{8}$

    C. $\frac{5}{12}$

    D. $\frac{5}{7}$

12. A bag contains 5 red marbles, 7 green marbles, and 4 blue marbles. Jeannie picked a green marble from the bag without looking. She did not put it back in the bag. Then Eva picked a marble from the bag. What is the probability that Eva's marble is red?

    A. $\frac{5}{16}$

    B. $\frac{1}{3}$

    C. $\frac{2}{5}$

    D. 1

13. Nick is trying to determine the favorite food of students in his school. Which sample below would be a random sample?

    A. all of Nick's friends

    B. every third student entering the school in the morning

    C. the students who are on the bus with Nick

    D. the students in Nick's math class

14. Bella plans to study the amount of time students in her school spend on homework. Which sample below is biased?

    A. a sample consisting of students in the honors classes

    B. a sample consisting of every fifth student who enters the cafeteria during lunchtime

    C. a sample consisting of one student randomly chosen from every homeroom

    D. a sample consisting of every tenth student from an alphabetical list of all students in the school

15. Mary drives to work every morning. The list below shows the time, in minutes, it took her to drive to work on 11 different days.

    21, 22, 22, 23, 24, 25, 25, 26, 26, 35, 38

    What is the lower quartile of the data set?

    A. 22 minutes

    B. 24 minutes

    C. 25 minutes

    D. 30 minutes

234

16. The box-and-whisker plot below shows the number of counties in each state in the United States.

Number of Counties

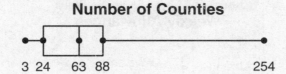

3 24   63 88   254

The box in the box-and-whisker plot is very short. The line connecting the box to the maximum value is long. What does this indicate about the data used to make the plot?

A. The data are clustered evenly within a small range.

B. Three quarters of states have 88 or fewer counties.

C. Data points are missing from the plot.

D. The data are evenly spread out throughout the range.

Directions: Use the following information to answer questions 17 through 19.

The students in Dwayne's class have taken eight tests this year. He wants to determine the mean, median, mode, and range of the scores of all the students. He selected two different samples. The following list shows Dwayne's first sample, which consists of all the test scores that he and his friend received.
92, 88, 85, 89, 95, 89, 87, 96, 89, 94, 96, 85, 99, 91, 83, 98

The following list shows the second sample, which consists of one random test score from each of the 18 students in his class.
76, 88, 81, 84, 94, 72, 80, 87, 75, 94, 79, 100, 82, 91, 84, 70, 84, 91

17. Find the mean, median, mode, and range of the test scores from Dwayne's first sample.

mean: _____ median: _____

mode: _____ range: _____

18. Find the mean, median, mode, and range of the test scores from Dwayne's second sample.

mean: _____ median: _____

mode: _____ range: _____

19. Which sample better represents the mean, median, mode, and range of all of the students' test scores? Explain.

_____

_____

20. The following dot plot represents the number of dollars in allowance that Sydney's classmates receive each week.

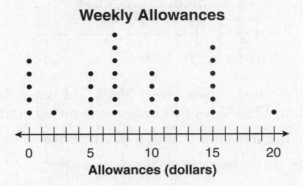

What is the mode of the weekly allowances? _____

Are there any outliers? Explain your answer.

_____

_____

What is the median weekly allowance?

_____

Explain how you found the median weekly allowance.

_____

_____

21. Parker is making an ice cream sundae. The sign shows the choices of ice creams, syrups, and toppings.

Ice Cream: vanilla or chocolate
Syrup: butterscotch, strawberry, or hot fudge
Topping: nuts or sprinkles

Draw a tree diagram to show all the choices Parker has for his ice cream sundae.

Directions: Use the data in the table to answer questions 22 through 24.

Day	Sunday	Monday	Tuesday	Wednesday	Thursday	Friday	Saturday
High Temperature	65°F	72°F	66°F	70°F	68°F	74°F	75°F

22. Display the data in the table in a dot plot on the number line below.

High Temperature (°F)

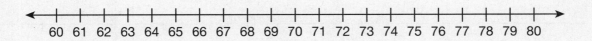

23. Display the data in the table in a box-and-whisker plot on the number line below.

High Temperature (°F)

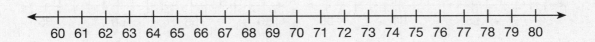

24. Find the mean absolute deviation of the data to the nearest tenth. Does the value of the mean absolute deviation indicate that the data are spread out or close together? Explain your answer.

_____

_____

25. You want to find out which movie showing in your town is most popular this week. You survey teenagers who are leaving one movie theater.

Explain why the sample is biased.

_____

_____

Explain how you can make the sample less biased.

_____

_____

26. Evaluate whether the measure of central tendency from the sample described below can be applied to the entire population. If it cannot be applied to the entire population, describe how the measure would differ between the sample and the population.

    sample: the average price of a meal at the most expensive restaurant in a city

    population: the average price of a meal at all restaurants in that city

    _____

    _____

    _____

27. Kristi and Dan are training for an upcoming swim meet. The dot plots show their practice times in seconds for the 50-meter freestyle race.

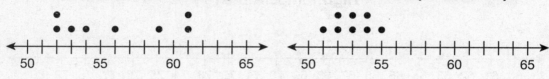

    Kristi 52, 52, 53, 54, 56, 59, 61, 61

    Dan 51, 52, 52, 53, 53, 54, 54, 55

    Part A
    Find the mean and the mean absolute deviation of Kristi's times.

    mean: _____   mean absolute deviation: _____

    Part B
    Find the mean and the mean absolute deviation of Dan's times.

    mean: _____   mean absolute deviation: _____

    Part C
    How much greater is the mean time for Kristi than the mean time for Dan?

    _____

    Part D
    About how many times greater is the mean absolute deviation for Kristi's times than for Dan's?

    _____

Math Tool: Coordinate Grids

Math Tool: Coordinate Grid

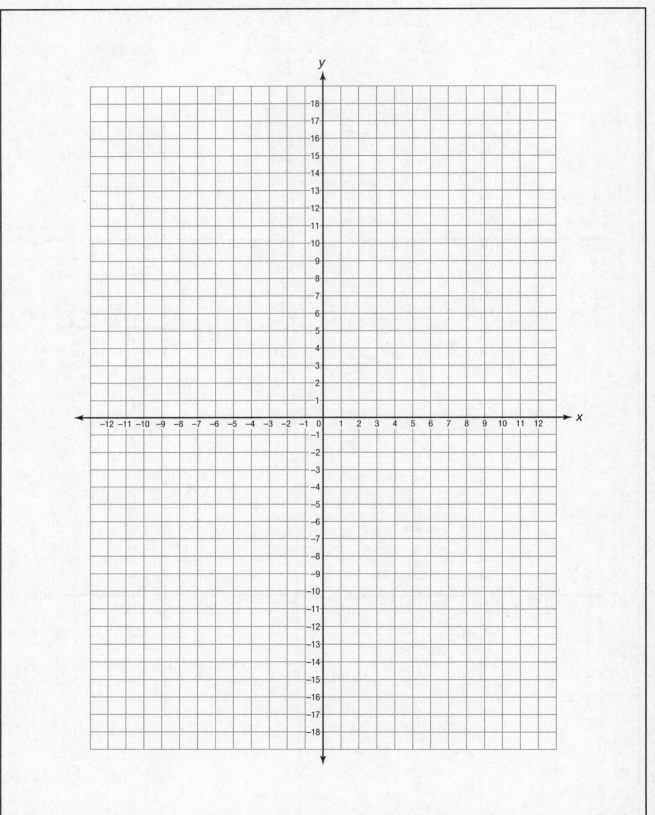